REALTOR EDGE

Mastering Trust, Clarity & Consistency in the Age of AI

By Stanley King & Chase Powell

A practical guide to building durable trust in modern real estate—without becoming performative, without chasing trends, and without losing your identity.

therealtorlab.com

Realtor Edge

ISBN: 978-1-960410-22-1

Printed in the United States of America

First Edition

Author Website: https://therealtorlab.com/edge-book

Table of Contents

FOREWORD

Stanley King

I'll never forget walking into the office one Tuesday morning in 2008.

The market had already started cracking. Listings were sitting. Closings were slowing. And when I stepped through the door, I saw something that stopped me cold: three agents sitting in cubes, staring at their screens, genuinely stressed—and genuinely confused about why their phones weren't ringing.

These weren't bad people. They weren't lazy. One of them had been in the business longer than I had. Another had just gotten licensed and was hungry to work. But they all had the same problem, and it took me a minute to name it:

Nobody knew they were real estate agents.

They had licenses. They had business cards somewhere in a drawer. But they weren't telling anybody. They didn't understand how to post on social media. They didn't know how to send an email that felt useful instead of salesy. They didn't even realize they were invisible—because in their minds, they were working hard.

That's when I started using a phrase I've never been able to shake: the Secret Agent problem.

A Secret Agent is someone who does great work—but does it so quietly that when opportunity shows up, they're not the name anyone remembers.

They're competent. They're ethical. They care. But they're not recalled.

And in real estate, if you're not recalled, you don't exist.

The belief that breaks good agents

I've been in this business for over two decades. I've sold more than 1,500 homes. I've worked through boom markets and the Great Recession. I've represented first-time buyers, luxury sellers, investors, families in crisis, and people starting over after loss. I've seen what works—and I've watched what fails.

And here's the hardest truth I know:

Most agents who struggle don't struggle because they're bad at real estate.

They struggle because they're invisible at the moment someone decides to move.

That invisibility isn't personal. It's structural. The world changed. Proximity used to do the heavy lifting. If you lived in the same town, went to the same school events, shopped at the same grocery store—people remembered you. Your presence stayed warm just by being around.

But today, your clients are surrounded by noise, options, and distraction. Even if they like you, there may be nothing in their daily life that consistently reminds them what you do, why you're good at it, or that you're even still active.

So when timing shows up—when the lease ends, the baby arrives, the transfer happens, the parent needs to downsize—they don't always think of you first. Not because you were rejected. Because you were forgotten.

And I've watched too many capable agents lose to that forgetting.

Why I care about this so much

Real estate isn't just a job to me. It's a calling.

I know that sounds old-fashioned, but I mean it. Housing is one of the most basic human needs—food, shelter, security, a place to recover and build. When you help someone find the right home, you're not just closing a transaction. You're changing their daily life. You're giving their kids stability. You're helping them start over, or scale up, or finally feel settled.

I've always believed that if you treat people like family—if you genuinely care about getting them into the right place, not just a place—they feel it. And when they feel it, they trust you. And when they trust you, they refer you. And when they refer you, your business becomes durable.

That's how I've built my career. Relationships. Standards. Showing up. Staying consistent. Treating every client like I'd treat my own kids if they were buying a home.

But here's the part that frustrates me:

I've watched agents who would be great at this—agents I'd trust with my own family—get passed over because they stayed quiet.

They assumed their work would speak for itself. They assumed reputation would carry them. They assumed that being good was enough.

And in the old world, maybe it was.

But that world is gone.

Visibility isn't ego. It's responsibility.

A lot of good agents resist visibility because they associate it with hype. I get it. This industry has earned that suspicion. There's a lot of performance. A lot of manufactured urgency. A lot of people saying things that sound impressive but feel hollow.

But visibility is not the same thing as noise.

Visibility, done right, is simply professional clarity. It's making sure that when someone in your sphere is ready to move, they don't have to work hard to remember you. They already know you're active. They already know you're competent. They already trust the way you think—because they've seen it.

You don't have to be loud. You don't have to perform. You don't have to become someone you're not.

You just have to be present enough that trust can form before the decision moment arrives.

That's not ego.

That's stewardship.

Why this book exists

When Chase and I started talking, it became obvious we were circling the same problem from different sides.

I've spent my career in the field—through luxury markets, land deals, foreclosures, flips, probate, and everything in between. I know what holds up when the market turns. I know what clients feel when they're scared. I know what separates agents who build durable businesses from agents who cycle through chaos every few years.

Chase has spent his career studying systems—how visibility actually works, how trust gets reinforced digitally, and how technology can amplify what's already there without diluting it.

Together, we realized something important:

This isn't a book about marketing tricks.

This is a book about understanding how clients actually choose, why trust forms—or breaks, and how to remain relevant without becoming performative.

And yes—AI is part of this story. But not in the way most people talk about it.

AI doesn't create trust. It can't replace experience. It can't supply discernment. What it can do, when used correctly, is protect consistency. It can reduce friction. It can help you stay present when time and energy get tight.

Used well, it's an engine of continuity. Used poorly, it accelerates sameness and erodes credibility at scale.

That line matters. This book draws it clearly.

What this book is—and what it isn't

This is not a book of shortcuts.

It won't promise overnight success. It won't tell you to "post more" just to post. It won't treat AI like magic. It won't turn you into someone you're not.

If you're looking for hacks, you'll be disappointed.

But if you're looking to build something durable—something that works when you're busy, something that survives when the market turns, something that lets you be yourself while still being seen—you're in the right place.

One more thing

You'll notice we reference tools and resources throughout the book. That's intentional. The concepts belong in these pages. The how-to belongs somewhere you can actually use it—checklists, audits, generators, templates.

When a tool is mentioned, you'll be able to find it at therealtorlab.com/edge.

Think of it as optional support, not an obstacle course. Use what helps. Ignore what doesn't.

My hope for you

My hope is that this book gives you language for what the best agents have always done.

It helps you make your competence legible.

It helps you stop disappearing.

It helps you show up in a way that feels like stewardship, not self-promotion.

Because real estate can change lives.

But only if people remember to call you.

— Stanley King Las Vegas, Nevada

FOREWORD

Chase Powell

I got my real estate license in 2003.

That detail matters because it means I didn't start as "a marketing guy" trying to sell services to an industry I didn't understand. I started as an agent, with the same pressure every agent feels: produce, perform, and somehow stand out in a crowded market.

And early on, I noticed something most people were slow to admit at the time: buyer behavior was changing faster than the industry was changing.

People were starting their journey online. They were researching quietly. They were forming trust impressions before they ever spoke to an agent. The game was shifting from purely relationship proximity to visibility plus credibility—and most agents were unequipped for that shift.

I didn't respond by chasing tricks. I responded by studying the machinery.

I went deep into websites, search behavior, content strategy, and the early foundations of what would later become mainstream digital marketing. I learned the hard truth that still applies today: you don't win online by "having a website." You win by building the kind of presence that matches how people actually search, evaluate, and decide.

That system-first thinking eventually became the foundation for my business work—building and advising on digital infrastructure that produces trust, not just traffic.

Why this book, why now

Over the past decade, I've designed marketing systems across multiple industries. I've seen what scales and what breaks. I've helped clients triple their business simply by sequencing the right systems and pipelines.

But here's what I've also seen:

Excellent agents getting overlooked—not because they aren't competent, but because their competence isn't consistently visible at the moment of decision.

They're doing the work. They're serving clients well. They're holding standards. But when their sphere is ready to move, the call goes to someone else. Someone louder. Someone more present. Someone who stayed top-of-mind through consistency, even if they weren't actually better.

That gap—between being good and being chosen—is what this book closes.

Where Stan and I connect

Stan brings decades of field experience. He's closed deals, weathered crises, and served families through their hardest and best moments. He knows what holds up when everything else falls apart.

I bring systems thinking, technology strategy, and the digital fluency that's become essential in the modern marketplace.

Together, we realized this book needed both perspectives:

- Stan's voice grounds the principles in human reality

- My voice translates those principles into repeatable systems

Sometimes you'll hear Stan's field-tested wisdom. Sometimes you'll hear my frameworks and digital strategies. Often, we'll trade off within a section. This dual perspective is deliberate: durable trust is built at the intersection of timeless human skill and modern leverage.

What makes this book different

This isn't another tactics book that will be outdated in six months.

This book is built on frameworks that survive platform changes, algorithm shifts, and technology cycles. We focus on mental models—how trust actually forms, why consistency compounds, what makes expertise legible—because those truths don't expire.

Yes, we talk about AI. But not as a magic solution. As an engine that can protect your consistency without flattening your identity—if you use it with judgment.

Yes, we talk about video, social media, and email. But not as separate tactics. As integrated channels that serve a single purpose: making your competence easier to recognize.

And yes, we give you implementation support. But we've separated the concepts (which belong in this book) from the tools (which live at therealtorlab.com/edge) so the book stays readable and durable even as specific platforms evolve.

My promise to you

If you take the ideas in this book seriously, one thing will happen:

You will stop being invisible.

Not because you become louder than everyone else—but because you learn how to make what you already have—your competence, your standards, your judgment—easy to see.

And when people can see it, they choose you.

Not because you convinced them.

Because you made the right choice obvious.

A final thought before we begin

Real estate is still one of the most personal businesses in the world.

People don't hire agents because they need access to listings. They hire agents because they're trying to make a decision that feels risky—and they want someone who can make it feel safer.

That's you.

But only if they can see you.

Let's make sure they can.

— Chase Powell Lubbock, Texas

INTRODUCTION
The Realtor Edge

There's a moment that happens in almost every real estate transaction—a moment most clients never see, and most agents never talk about.

It's the moment when everything could go sideways.

The appraisal comes in low. The inspection uncovers something unexpected. The buyer gets cold feet. The seller panics. The timeline compresses. Emotions spike. And suddenly, what looked like a smooth process feels like it's about to crack.

That's the moment your value shows up.

Not in the marketing. Not in the photos. Not in the listing description.

In the calm you bring to chaos.

In your ability to say, "Here's what this means. Here's what matters. Here's what we do next."

In the way you turn a crisis into a decision.

That moment—that skill—is what clients are actually paying for. They're not hiring you to unlock doors or submit paperwork. They're hiring you to stand between them and uncertainty and make a high-stakes decision feel navigable.

That has always been true.

What has changed is the environment trust has to survive in.

The modern problem, in one line

Your clients are surrounded by options, information, opinions, and noise.

Their attention is fractured by design. They have headlines, interest rates, rumors, Zillow estimates, group chats, and social media feeds all competing to tell them what's "about to happen." Even when they like you—even when they trust you personally—there may be nothing in their day-to-day life that consistently reminds them what you do, what you know, and why you're the safest person to call.

That's the shift:

Competence no longer makes itself visible.

You can be experienced, ethical, and hardworking—and still lose business for a simple reason: you weren't the name that surfaced at the decision moment.

Not because you were rejected.

Because you were forgotten.

The gap this book closes

This book is about closing the gap between being competent and being chosen.

It's about making your competence legible—easy to recognize, easy to trust, and easy to recall when someone finally says, "We need an agent."

And here's what that requires:

It requires understanding how trust actually forms in a noisy world. It requires knowing why expertise compounds—and why silence erodes it. It requires building systems that keep you present even when your schedule is full. It requires using AI intelligently—as an engine, not a shortcut. It requires making your presence confirm what your best clients already know.

That sequence is intentional.

Because if you start with tactics—post more, run ads, try video—you end up chasing platforms instead of building trust. And trust is what survives when everything else changes.

A story that makes this real

Most agents think their business comes from what they do during a transaction.

Stan's business comes from what happens between transactions.

The monthly market updates that explain what rising inventory actually means for buyers in specific neighborhoods. The posts that break down inspection reports in language normal humans can understand. The emails that share local insights—not because there's a listing to promote, but because the information is genuinely useful.

He calls them "trust deposits."

Each one is small. Individually, none of them feel significant. A three-minute video walking through a neighborhood. A text checking in after a closing. An email explaining why interest rate headlines don't tell the whole story. A social media post about a local restaurant opening.

None of these things close a deal today.

But over twenty-two years, those deposits have created something that can't be manufactured overnight, can't be purchased with ad spend, and can't be replicated by a competitor who just entered the market: a reputation for being the agent who makes hard decisions feel manageable.

When timing activates—when the baby arrives, when the job transfer comes through, when the parent needs to downsize, when the lease renewal notice triggers a decision—people don't launch a research project.

They remember.

They remember the agent who explained the appraisal process when they weren't even in the market. They remember the agent whose posts helped them understand their neighborhood's value trends. They remember the agent who stayed present, stayed useful, and stayed calm when everyone else was creating urgency.

And here's what most agents miss: that memory isn't created in a single moment. It's accumulated through dozens of micro-interactions that individually feel insignificant but collectively create gravity.

One market update doesn't make you memorable. Fifty-two of them, delivered consistently, make you inevitable.

One helpful post doesn't build trust. A year of helpful posts makes you the obvious call.

One check-in doesn't create loyalty. A pattern of checking in—without asking for business—creates a relationship that survives market shifts, platform changes, and competitors who try to buy attention with bigger budgets.

This is the compounding advantage.

Not louder. Not flashier. Not more aggressive.

Just more consistent. More useful. More present.

And when you compound trust over years, something remarkable happens: you stop chasing business. Business starts finding you.

Not because you convinced anyone. Because you made the right choice obvious before they ever needed to choose.

That's what this book protects.

The practices, systems, and mindsets that let you build trust deposits when clients aren't buying—so when they are, you're the only name they consider.

Most agents compete on the transaction.

Market leaders compete on the years between transactions.

What "The Realtor Edge" actually means

The Realtor Edge is not charisma.

It's not volume. It's not "being good at social media."

The Realtor Edge is the ability to remain relevant, credible, and chosen in a market that is noisy, uncertain, and constantly changing.

You build it through five levers:

1. Trust as a discipline — understanding how trust forms and what breaks it
2. Expertise that compounds — teaching in a way that reduces uncertainty
3. Systems that create consistency — so your presence doesn't depend on your mood or schedule
4. AI used with judgment — not as a shortcut, but as an engine for standards and continuity

5. Legibility — translating invisible competence into visible proof

The sequence matters.

Trust comes first. Systems make trust repeatable. AI comes later—not because AI isn't powerful, but because power without clarity produces noise.

What this book is—and what it isn't

This is a do-it-yourself book, but not a do-everything-yourself book.

Inside these pages, we'll stay at the level that actually changes outcomes: principles, frameworks, and mental models. We'll show you what matters, why it matters, and how to think about it the way market leaders think about it.

We won't drown you in step-by-step tactics that become outdated next quarter. We won't ask you to live online. We won't treat AI like magic.

AI can help you move faster and stay consistent. It can reduce friction and protect your standards. But it cannot supply judgment, integrity, or taste.

Used well, AI helps you communicate with more clarity and consistency. Used poorly, it accelerates generic output and erodes credibility at scale.

We will draw that line clearly.

The promise (and the structure)

If you take the ideas in this book seriously, one thing will happen:

You will stop being invisible.

Not because you become louder than everyone else—but because you learn how to make what you already have—your competence, your standards, your judgment—easy to see.

Here's how we'll get there:

Part I establishes why trust is the new currency—and why competence without signal is a liability.

Part II shows you how expertise compounds when you teach instead of just transact.

Part III builds the systems that keep you present even when life gets full.

Part IV positions AI correctly—as an engine, not an identity.

Part V makes trust legible—so when clients check, what they see confirms what your best clients already know.

Part VI protects durability—so you can adapt without fragmenting.

And in the conclusion, we'll tie it all together into a single operating system you can run for the next decade, regardless of what changes around you.

A note on tools and resources

Throughout this book, you'll see references to worksheets, audits, generators, and templates.

We separated them intentionally.

The concepts belong here. The implementation tools belong where you can actually use them—organized, accessible, and optional.

When a resource is mentioned, you'll find it at therealtorlab.com/edge.

Think of it as a companion vault, not a requirement. Use what helps. Ignore what doesn't.

The book stands on its own.

The tools simply make execution easier.

Additional thoughts before we begin

Real estate is still one of the most personal businesses in the world.

People don't hire agents because they need access to listings. They hire agents because they're trying to make a decision that feels risky—and they want someone who can make it feel safer.

That's you.

But only if they can see you.

Let's make sure they can.

The Real Estate Dream: Freedom, Income, and Impact

Before we dive into trust, systems, and AI, let's be direct about something most real estate books tiptoe around: why agents actually choose this profession.

Yes, you care about helping people. Yes, housing matters. Yes, guiding families through major life transitions is meaningful work.

But there's another truth that deserves to be said clearly:

Real estate offers one of the last true paths to financial freedom and entrepreneurial independence in America.

The barrier to entry is remarkably low. Three to six months of focused study. A licensing exam. And suddenly, you have access to an industry where individuals routinely earn six figures—and the top performers earn seven.

There's no corporate ladder to climb. No waiting for promotions. No arbitrary salary cap determined by someone else's budget cycle.

In real estate, your income is determined by one thing: your ability to build trust, close transactions, and create a business that compounds over time.

This is raw capitalism at its finest. You can work as hard as you want—or as little. You can choose your own schedule, your own clients, your own standards. You can build a business that generates substantial income while giving you time freedom most nine-to-five workers will never experience.

Want to take Fridays off? Build a business that allows it. Want to work 60-hour weeks and bank $300,000 in a year? The market will reward that effort. Want to sell $200 million in real estate and retire after five years? People have done it.

The freedom is real.

You're not an employee. You're an independent contractor. A true entrepreneur. You own your book of business. You control your pipeline. You decide when to show up and what to prioritize.

But here's the uncomfortable part most agents discover too late:

That same freedom destroys the unprepared.

Without systems, you drown in chaos. Without trust, you're invisible. Without consistency, your pipeline evaporates every time you get busy. Without positioning, you compete on price instead of value.

The agents who thrive aren't just good at transactions. They're good at building businesses that don't require constant hustle to survive.

They understand that real estate success isn't about working harder than everyone else. It's about building the kind of trust that makes people choose you, refer you, and come back to you—even when the market shifts, even when competitors offer lower fees, even when you're too busy to chase leads.

That's what this book protects.

The dream is real: financial independence, time freedom, the ability to serve families while building wealth.

But the dream only survives if you build the foundation that makes it durable.

The systems we're about to show you aren't just about closing more deals. They're about creating a business that generates income, freedom, and impact—without burning you out or fragmenting your identity in the process.

So yes—this book is about trust, expertise, and AI.

But it's also about building the kind of life most people only talk about wanting.

Let's make sure you actually get there.

Let's begin.

PART

I

TRUST:
THE NEW CURRENCY

Trust Is No Longer Assumed

Stan has a rule he's followed for twenty-two years:

Always answer your phone.

It sounds simple. Almost quaint. But he means it literally—if a client calls, he picks up. If he's in a meeting, he steps out. If he's with his family, he excuses himself for sixty seconds. If he's on another call, he calls back within minutes.

Not because he's obsessed with being available.

Because he knows what happens if he doesn't.

The fifteen-minute window

Here's what Stan learned early: when someone decides they're ready to move, they don't create a list of ten agents and methodically interview each one.

They think of a name.

They call.

And if that person doesn't answer—or takes too long to respond—they call the next name.

The decision window is fifteen minutes. Maybe less.

It's not that clients are impatient or unreasonable. It's that the need feels urgent. The lease is ending. The job transfer just came through. The baby is coming. The parent just fell. Something shifted, and now the question is real:

"We need to move. Who do we call?"

And in that moment, they don't want to wait. They want to feel like the decision is in motion. They want relief from the uncertainty of not knowing what happens next.

So they call the agent whose name surfaced first.

And if that agent is unavailable, the client doesn't pause and wait. They move to the next name—because the need is still there, and someone else will answer.

That's the reality of modern real estate.

Trust isn't assumed. It's activated.

And if you're not reachable at the moment of activation, you've already lost.

The quiet shift that changed everything

There was a time when trust came preloaded.

Proximity did a lot of the work. People saw you at school functions, in the neighborhood, at the coffee shop, at church. Your sphere knew your rhythm. Even if they weren't actively thinking about buying or selling, your presence stayed warm in their mind simply because life brought you into view.

You didn't have to remind people you were a Realtor. They just knew.

But that world isn't gone—it's just no longer the whole world.

Today, your clients live in an environment designed to fracture attention. Even if they like you, even if they trust you personally, there may be nothing in their daily routine that consistently reinforces what you do, why you're good at it, or that you're even still active.

Meanwhile, they're being exposed to other agents—some in their feed, some on signs, some in group chats, some through friends. Those agents may not be better. They may not be as competent. But they are present.

And presence is interpreted as relevance.

This is the shift:

Trust is no longer assumed through proximity. Trust is earned through reinforcement.

When uncertainty rises, people buy stability

Most agents think they're competing on service, or expertise, or market knowledge.

In reality, you're competing on certainty.

Real estate decisions are emotionally expensive because the stakes are high and the variables feel unstable. Clients aren't just wondering about price—they're wondering about timing, risk, regret, and exposure.

The questions they carry are usually some version of:

- Are we buying at the top?
- Will we get trapped with the wrong house?
- What if we list and can't find a replacement?
- What if the inspection uncovers something serious?
- What if we wait and lose momentum—or move and feel foolish later?

When uncertainty rises, humans don't primarily seek more information. They seek someone who can interpret information without inflaming fear.

They seek stability.

They seek someone who can reduce ambiguity, explain tradeoffs, and help them feel less exposed.

That is what trust actually is in this context:

A reduction of perceived risk.

Stan's story: Trust under pressure

During the Great Recession, Las Vegas was one of the hardest-hit markets in the country. Home values dropped as much as 84% in some areas. Construction stopped. Jobs evaporated. Foreclosures flooded the market.

Stan was hired by banks to handle properties where the former owners were still living inside—people who had stopped paying their mortgages, often because they were underwater by hundreds of thousands of dollars.

Most agents avoided that work. It was uncomfortable. You had to knock on someone's door and essentially tell them their hardest day had arrived.

But Stan didn't see it that way.

He saw families in crisis. People who were scared, embarrassed, and often had no idea what came next.

So he showed up differently.

He knocked on the door with respect. He introduced himself calmly. He didn't treat them like a problem to remove—he treated them like people trying to navigate an impossible situation.

And instead of just delivering bad news, he worked with the bank to get them something they didn't expect: a cash settlement—sometimes $5,000, sometimes $10,000—enough to cover their move and leave with dignity.

Most of those people were devastated. They were losing what they thought was the foundation of their future.

But they were also relieved.

Because Stan gave them clarity. He gave them a path forward. He gave them respect when they felt powerless.

And here's the part that matters:

Years later, some of those people called Stan back.

They were ready to buy again. Their credit had recovered. Their lives had stabilized.

And they wanted him to represent them.

Not because he sold them a house the first time. Not because he gave them a deal.

Because he treated them like family when they had nothing to offer him.

That's trust.

Trust isn't built in the easy moments. It's built in the hard ones.

Trust is what happens when you stay calm when everyone else is panicking. When you tell the truth without dramatizing it. When you frame the situation clearly enough that people can make a decision instead of spiraling.

And trust—real trust—compounds.

It doesn't just close one deal. It creates referrals. It creates a reputation. It creates the kind of business that doesn't reset every time the market turns.

Field-Tested Principle

"Trust isn't built by avoiding hard moments. It's built by staying steady inside them."

Why "trust" is the new currency

In a trust economy, clients don't always choose the "best" agent in an objective sense—because they don't know how to measure "best" before they hire.

They choose the agent who feels like the safest decision.

Safety, in this context, isn't about being conservative. It's about feeling guided.

Safety looks like:

- Clear explanations, without jargon
- Calm tone, even when the news is complicated
- Specificity instead of generic reassurance
- Proof that you're active and current, not stale or sporadic
- Consistency that signals reliability

None of this requires you to be loud.

But it does require you to be legible.

This is where many great agents struggle: their competence is real, but it is not easily recognizable unless someone is already inside the relationship. They are excellent in private and nearly invisible in public.

And in a market where trust must be reinforced, invisibility is expensive.

The uncomfortable truth about disappearing

Here's the part most agents don't want to hear—not because it's harsh, but because it feels impersonal:

If someone doesn't see you for long enough, your position in their mind decays.

That's not a comment on your character. It's how attention works.

The brain updates its "most relevant" list based on what it sees repeatedly. The agent who shows up with consistency becomes easier to recall. The agent who shows up occasionally—even if they are brilliant—becomes a "maybe."

This is why agents who maintain a steady, non-salesy presence often win opportunities they didn't even know existed. They are not necessarily better. They are simply easier to remember at the moment a decision is made.

And here's the painful irony:

The agents who disappear most often are the ones who are busiest.

They're executing. They're serving clients. They're doing the work. But while they're heads-down in transactions, their visibility goes dark. Their sphere stops hearing from them. Their social media goes quiet. Their email list forgets they exist.

And when the busy season ends, they realize their pipeline feels thinner than it should—because their relevance cooled while they were working.

That's the trust drop-off problem.

And it's solvable.

Two kinds of trust—and why one compounds

Most people think trust is emotional: "I trust you because I like you."

That matters. But in real estate, especially in a complex market, the kind of trust that compounds is competence trust:

"I trust you because you help me understand what's happening, and I believe you'll handle the hard parts well."

Competence trust is built through how you communicate—how you frame risk, explain tradeoffs, and stay calm in complexity.

It's built by showing your thinking, not by showing off.

The agents who become "the obvious call" aren't always the most charismatic. They're the ones who create the feeling of stability. They reduce uncertainty when everyone else adds to it.

They make hard things feel solvable.

Coaching Moment

Stop asking: "How do I get more leads?"

Start asking: "How do I become the person people feel safer with?"

That's the whole game.

The question that changes the rest of this book

This section isn't asking you to become a different person. It's asking you to see the game clearly.

Trust is currency. And currency has to be earned and maintained.

So here's the question to hold in your mind as we move forward:

If someone in your sphere needed an agent this week, would they think of you first—or would they have to remember you?

If the answer is "they'd have to remember," don't treat it like a condemnation.

Treat it like a diagnosis.

Because the solution is not to be louder. It's to be clearer, more consistent, and more legible—so trust can form before the decision moment arrives.

That's where we go next.

The Friend, Not the Agent

Stan has a confession that might sound unprofessional at first:

He hates being someone's "agent."

Not because he doesn't love real estate. He's been doing this for over two decades. He's closed more than 1,500 transactions. He's represented luxury buyers, land investors, foreclosure clients, and families in crisis.

But when he walks into a relationship where the client sees him as "their Realtor"—as a service provider, a vendor, someone on the other side of a transactional line—the entire experience feels wrong.

He doesn't want to be hired. He wants to be trusted.

He doesn't want to work with clients. He wants to work with friends.

And that distinction changes everything.

The laugh test

Stan has a technique he's used for his entire career. It's so simple it sounds almost trivial. But it's also the foundation of every strong relationship he's ever built in real estate.

When he meets someone—whether it's a listing appointment, a buyer consultation, or just a conversation at a neighborhood event—he does one thing immediately:

He makes them laugh.

Not with a canned joke. Not with rehearsed charisma. Just genuine human connection.

He finds commonality. He says something real. He drops the professional façade and talks like an actual person instead of someone performing the role of "real estate expert."

And as soon as they laugh—as soon as that guard drops and the interaction feels easy—he knows something important:

This person isn't going to be difficult. They're not an asshole. And more importantly, rapport has been established.

They're not strangers anymore. They're not on opposite sides of a transaction.

They're friends.

And once you're friends, everything changes.

Why transactions fail when relationships don't exist

Stan doesn't just prefer working with friends. He's learned over twenty-two years that it's the only way the work stays sustainable.

When a client sees you as "their agent," the relationship has a ceiling. They expect you to perform. You expect them to comply. The dynamic is transactional, which means it's fragile.

And real estate transactions are messy.

Appraisals come in low. Inspections uncover problems. Timelines compress. Emotions spike. Sellers panic. Buyers get cold feet.

In those moments, if the relationship is purely transactional, everything fractures.

The client questions your advice. You feel undervalued. Tension builds. The deal becomes a grind instead of a collaboration.

But when you're friends?

When you've already established that you genuinely care about them—not just about closing the deal—the hard moments don't break the relationship. They deepen it.

They trust your judgment because they know you're not just trying to make a commission. You trust their intentions because you've already built rapport. The transaction becomes a shared project instead of a battle.

The thing AI will never understand

Here's the part that matters in an era obsessed with automation, efficiency, and AI:

No algorithm will ever replace the moment when two people laugh together and become friends.

AI can draft emails. It can generate social media posts. It can analyze market data and produce reports.

But it cannot create the human connection that happens when you drop the performance, find commonality, and treat someone like a person instead of a prospect.

That moment—when the professional distance dissolves and you're just two people who understand each other—is where trust is actually built.

Not through competence. Not through credentials. Not through marketing.

Through humanity.

Treating clients like family

Stan doesn't just say he treats clients like friends. He treats them like family.

When a client calls with a crisis, he doesn't think, "This is interrupting my day." He thinks, "This is someone I care about, and they need help."

When a deal gets complicated, he doesn't calculate how much time it's costing him. He asks, "What would I do if this were my own kid buying their first house?"

That filter—"Would I advise my own family to do this?"—guides every decision he makes.

And clients feel it.

They feel the difference between someone who's performing the role of "helpful agent" and someone who genuinely cares whether they end up in the right house.

The brutal honesty this allows

Here's the surprising advantage of the friend-not-agent approach:

It gives you permission to tell the truth.

When you're just "the agent," clients expect you to be diplomatic, cautious, and agreeable. They expect you to manage their emotions and avoid conflict.

But when you're a friend, you can be honest in ways that would feel harsh coming from a vendor.

Stan has told buyers, "This house is wrong for you. I know you love it, but here's why you'll regret it in two years."

He's told sellers, "Your price expectations are unrealistic, and here's the data that proves it. We need to adjust or you're going to sit on the market and lose momentum."

Those conversations don't work when the relationship is transactional. The client hears criticism and shuts down.

But when the relationship is built on friendship—when the client knows you care about their outcome more than your commission—they listen. Because they trust that your honesty is coming from care, not convenience.

The business model this creates

Most agents think of their business in terms of "lead generation" and "client acquisition."

Stan thinks of his business differently:

He builds friendships. And friends refer friends.

When you close a transaction with someone who feels like they hired an agent, they might leave a review. They might send a thank-you note. But the relationship usually ends there.

When you close a transaction with someone who feels like they worked with a friend, something different happens:

They tell people.

Not because you asked for referrals. Not because you reminded them you're in real estate.

Because when someone asks, "Do you know a good Realtor?" their brain doesn't retrieve a business card. It retrieves a relationship.

And they say, "Yeah, call Stan. He's not just good—he actually cares. He treated us like family."

That's the compounding advantage.

You're not competing on marketing volume or ad spend or social media followers.

You're competing on depth of relationship—and relationships compound faster than visibility.

The positioning this requires

Here's the hard part:

You can't fake this.

Clients can tell the difference between someone performing friendliness and someone who actually operates from care.

The agent who says "I treat my clients like family" as a tagline sounds like everyone else.

The agent who actually treats clients like family—who answers the phone at 9pm, who doesn't calculate billable hours, who makes decisions based on "What would I want for my own kids?"—doesn't need a tagline. The positioning is self-evident.

This isn't about marketing. It's about identity.

It's about deciding whether you're building a business or building a life's work.

And if you choose the latter, the business takes care of itself.

Field-Tested Principle

"I don't work with clients. I work with friends. And friends don't forget you when they need help."

Why this matters before we talk about systems

The rest of this book will show you how to scale trust, use AI intelligently, and build systems that keep you visible even when you're busy.

But none of those tools matter if the foundation isn't right.

Because if you're trying to scale a transactional business—one built on "client acquisition" instead of relationship depth—you'll always be chasing the next lead, the next post, the next tactic.

But if you build from friendship first, the systems become amplifiers instead of replacements.

AI doesn't create the relationship. It protects the consistency that keeps the relationship warm.

Content doesn't manufacture trust. It reinforces trust that's already been built through human connection.

Systems don't replace care. They ensure care can scale without burning you out.

This is the sequence that matters:

First, become a friend. Then, use systems and AI to stay present as a friend—even when life gets full.

That's the only way this works long-term.

And it's the only way the business becomes sustainable, referral-driven, and deeply fulfilling instead of exhausting.

Now let's talk about what makes competence visible.

Competence Is Invisible Without Signal

Most agents don't have a competence problem.

They have a translation problem.

They can price a home with nuance. They can read a deal and sense where it's going to crack. They can negotiate without turning the room into a war zone. They can keep a transaction moving when the emotional temperature rises and everyone else starts making decisions out of fear.

But the market doesn't reward what you know.

It rewards what people can recognize.

And recognition depends on signals.

The problem with "excellent work"

Here's the paradox: the better you are, the less dramatic your work looks.

When an agent does their job well, the client experiences fewer surprises. The timeline feels smoother. The negotiation feels calmer. The little problems that would have become big problems get handled before they become visible.

Clients often walk away thinking, "That wasn't as bad as I expected."

Which is exactly what you want for them.

But it creates a challenge for you: the most valuable parts of your competence are the parts clients rarely see until they're already inside the relationship.

And in today's market, people choose before they can fully evaluate.

If your competence only becomes obvious after someone hires you, you'll always be fighting uphill against agents whose competence is easier to "feel" from the outside.

What clients can't measure, they replace with familiarity

Most clients aren't trying to be unfair. They're trying to make a safe decision with limited ability to evaluate outcomes in advance.

Before they hire you, they can't truly measure your negotiation skill. They can't observe your systems. They can't see the standards you hold when nobody is watching. They can't know whether you'll be calm during the inspection, or strategic during pricing, or steady when emotions spike.

So they substitute a simpler metric: familiarity.

"Do I feel like I know this person?" "Do they seem active?" "Do they seem current?" "Have I seen them explain things clearly before?" "Can I picture them guiding me through this?"

Those are not shallow questions. They're survival questions. They are the brain's shortcut for risk reduction.

The danger is that familiarity can be purchased cheaply—through volume, hype, or constant posting—while competence trust must be earned.

That's why the goal isn't to become more visible for visibility's sake.

The goal is to become visible in a way that proves competence.

That's what signals do.

What a signal really is

A signal is not a slogan. It's not a logo. It's not "Top Producer" pasted into a bio.

A signal is evidence.

It's something your market can see that indicates how you think, how you operate, and what working with you would feel like.

Signals reduce uncertainty because they show people the texture of your judgment.

The best signals do one of three things:

1. They create clarity.
2. They create specificity.
3. They create confidence through consistency.

Signal #1: Clarity

Clarity is one of the rarest commodities in real estate because most communication is either too vague to be helpful or too technical to be understood.

Clarity doesn't mean certainty. It means you can take messy information and turn it into something actionable.

Clients trust the agent who can say, calmly and honestly:

- "Here's what's true right now."

- "Here's what's noise."
- "Here's the tradeoff you're actually making."
- "Here's the risk, and here's how we reduce it."

When you communicate like that, you aren't just giving information. You're providing stability. You're helping people feel less exposed.

Consider the difference between a generic social media post and a clarity-driven one:

Generic post: "New listing! 3 bed, 2 bath in a great neighborhood. Call me for details!"

This tells your audience almost nothing about why the home matters or what action they should take.

Clarity-driven post: "This three-bedroom bungalow sits in the Jefferson school district, where home values have appreciated 8% year over year. It's priced at $420K, below the neighborhood median, and the seller is open to a contingency clause. If you're thinking about long-term value or need to time a sale and purchase, let's talk about how this fits your plans."

The latter frames data in a way that matters to your audience and invites a specific conversation.

That is competence trust being built in real time.

Signal #2: Specificity

Generic is the enemy of trust.

Most agents try to sound professional and end up sounding identical. They say the same phrases, share the same stock advice, post the same market stats without interpretation, and describe themselves in ways that could fit anyone.

Specificity is what breaks the pattern.

Specificity is when someone can look at you and think, "They're for people like me." Or, "They know my neighborhood." Or, "They've clearly done this exact kind of transaction."

This doesn't require you to niche into a microscopic box. It requires you to stop trying to be the right agent for everyone.

The market rewards the agent who feels like the best fit for a clear set of people and situations.

Specificity signals show up in simple places:

- The way you describe your market
- The examples you use
- The kinds of decisions you explain
- The problems you seem fluent in
- The language you use to frame tradeoffs

When you're specific, you're not excluding people. You're becoming believable.

Signal #3: Consistency

Consistency is the signal most agents underestimate, because it doesn't feel like "marketing." It feels like a discipline issue.

But your market reads consistency as reliability.

If you show up in bursts—high visibility when you need business, silence when you're busy—the message your audience receives is not, "They're busy."

The message is, "They are unpredictable."

And in a trust economy, unpredictable feels risky.

Consistency does not mean you post daily. It means you have a steady, minimum viable presence that doesn't disappear for months at a time. It means your sphere doesn't have to guess whether you're active. They can see it.

When you combine clarity, specificity, and consistency, something important happens:

You become easy to recall.

Not because you're loud, but because you're dependable.

Stan's story: "I label myself everywhere"

Stan has a habit that sounds almost obsessive until you understand why he does it.

Everywhere his name appears—email signatures, social media profiles, website headers, even casual text messages—he includes one phrase:

Stanley King, Realtor, Las Vegas

Not "Stanley King, Real Estate Professional." Not "Stanley King, Broker."

Realtor. Las Vegas.

Simple. Clear. Consistent.

Why?

Because he knows how people search. He knows how memory works. He knows that when someone hears his name in passing, or sees it in a group chat, or gets referred by a friend, they're going to do one thing:

They're going to Google him.

And when they do, he wants them to see the same signal everywhere: This person is a Realtor in Las Vegas.

Not once. Not sometimes. Everywhere.

It's not branding. It's cognitive reinforcement.

It makes him easier to find. Easier to remember. Easier to trust.

Because trust isn't built by cleverness. It's built by coherence.

Field-Tested Principle

"Your market doesn't reward creativity. It rewards coherence."

The legibility test

A simple test makes this real.

Imagine someone in your sphere hasn't spoken to you in six months. They're suddenly thinking about moving. They open their phone. They search your name.

What would they see that proves you're the right agent?

Not what you believe about yourself—what would they actually see?

Would they see calm explanations? Local interpretation? Evidence that you're active? A sense of how you guide decisions?

Or would they see a few generic posts that could belong to anyone?

That question isn't meant to shame you. It's meant to give you a clean measurement.

Because if competence is invisible without signal, the first move isn't "more content."

The first move is making sure the signals you do send are doing their job.

A better definition of marketing

When marketing is done poorly, it feels like persuasion.

When marketing is done well, it feels like service.

Here's the reframe that keeps you grounded:

Marketing is not convincing people you're good. Marketing is making it easy for the right people to recognize that you're already good.

Once you accept that, you stop chasing tactics.

You start building signals that reflect who you actually are—and that is what creates durable trust.

Coaching Moment

Your job is not to impress strangers.

Your job is to make your competence legible to the people who already trust you—so when timing activates, you're the obvious call.

In the next section, we'll talk about the agent this market punishes the most: the competent professional who stays hidden, not because they lack skill, but because visibility feels like ego.

We'll also show why that belief—while understandable—comes with a cost.

The "Secret Agent" Problem

A Secret Agent is not a beginner.

A Secret Agent is not lazy.

In many cases, a Secret Agent is the kind of agent you would want on your side if your deal had complications: thoughtful, ethical, detail-oriented, calm under pressure.

They're also the kind of agent who is most likely to feel disillusioned right now.

Because they're doing the work.

They're showing up for clients. They're learning the market. They're trying to operate with integrity. And yet they're watching other agents—sometimes agents they privately don't respect—get opportunities that seem disproportionate to their skill.

It can feel like the market is rewarding the wrong things.

In a way, it is.

But not because the market values hype over competence.

Because the market values what it can see.

Stan's story: Agents in the cube, freaking out

Stan remembers the moment he gave the problem a name.

It was 2008. The market had started cracking. Listings were sitting longer. Closings were slowing. And when he walked into the office one morning, he saw three agents sitting in their cubes, staring at their screens, genuinely stressed.

These weren't bad people. One of them had been in the business longer than Stan. Another had just gotten licensed and was hungry to work. The third was smart, professional, and capable.

But they all had the same problem:

Nobody knew they were real estate agents.

They had licenses. They had business cards somewhere in a drawer. They could talk intelligently about contracts, neighborhoods, and market conditions.

But they weren't telling anybody.

They didn't understand social media. They didn't know how to send an email that felt useful instead of salesy. They weren't posting. They weren't staying visible. They weren't reinforcing their presence in any consistent way.

And because of that, when people in their own circles needed an agent, those people called someone else.

Not because these agents were incompetent.

Because they were invisible.

Stan started calling them Secret Agents.

They had the license. They had the knowledge. They could do the work.

But they were operating so quietly that when opportunity showed up, they weren't the name anyone remembered.

How Secret Agents are created

Most Secret Agents became Secret Agents for reasons that make sense.

Some were taught that talking about yourself is arrogant. They learned that "real professionals" let their work speak. Some came from industries where visibility wasn't part of the job, and self-promotion was considered a red flag.

Some have watched marketing in this industry turn into performance and noise, and they want nothing to do with it.

Others are simply overwhelmed—busy with transactions, family, life—and visibility feels like one more burden they can't carry.

So they stay quiet. They wait for someone to ask. They assume their reputation will carry them.

And if we're being honest, in some markets and some eras, it did.

But in a trust economy, silence is not interpreted as virtue.

Silence is interpreted as absence.

The cost of invisibility is not just fewer leads

Invisibility doesn't only cost you volume. It costs you quality.

When you aren't legible, you don't consistently attract the clients who value what you do best. You get whatever filters through the cracks: last-minute chaos, price shoppers, people who found you accidentally, transactions that feel like emergencies rather than relationships.

Meanwhile, the clients who would appreciate your calm and your standards choose someone else—because someone else felt more present.

This is why Secret Agents often feel like they're always rebuilding.

Every quiet season resets their relevance. Every busy season makes them disappear again. Their business doesn't compound; it cycles.

And the cycling creates exhaustion.

The difference between humble and hidden

Humility is a strength. Being hidden is not.

A humble agent does not exaggerate. A hidden agent is not evaluated.

A humble agent keeps the focus on the client. A hidden agent doesn't get the call.

The distinction matters because Secret Agents often tell themselves a story that sounds noble, but functions as a defense:

"I'm not like those other agents."

Maybe you aren't. But the question isn't whether you're "like them."

The question is whether the people who need you can recognize you as the safe choice when timing activates.

If they can't, your standards don't get rewarded. They get overlooked.

Field-Tested Principle

"Being good in private doesn't help people who can't find you."

Minimum viable presence is not ego

The answer is not becoming loud.

The answer is becoming consistently present at a minimum viable level—enough that your sphere never has to wonder whether you're active, current, and competent.

Minimum viable presence can be remarkably simple.

It can be one market interpretation per week. It can be one clear explanation of a common question. It can be periodic proof that you are in the business, in the work, and in the market.

It doesn't require theatrics. It requires continuity.

This is the mindset shift Secret Agents need to make:

Visibility is not self-promotion. Visibility is client protection.

Because when people cannot see competence, they default to familiarity. And when they default to familiarity, they may choose an agent who is louder but less capable—which creates risk for them.

If you care about clients, you have to care about legibility.

Stan's story: "I'm not trying to be like, 'Hey, I'm Stan'"

Stan puts it this way:

"When I joined a country club, or a car club, or when my kids started school, I didn't go in thinking, 'I need to sell these people.' I went in because I wanted to connect. I wanted to know people. I wanted to be part of something."

"But I also didn't hide what I do."

"I'd introduce myself. I'd hand out my card. I'd say, 'If you ever need help with real estate, let me know.' That's it. I wasn't pitching. I wasn't performing. I was just making sure they knew."

"And you know what? A lot of those people became clients. Not because I pressured them. Because when they needed an agent, they remembered me. They knew I was active. They knew I cared. They knew I was real."

That's the whole strategy.

Not louder. Not performative. Just present.

The relief on the other side

Here's the good news: most Secret Agents don't need to become "better" at real estate.

They need to become easier to recall.

And that is a design problem, not a personality problem.

When you build a steady signal—clarity, specificity, consistency—your business begins to compound. You stop living at the mercy of timing. You stop feeling like you have to reintroduce yourself every time the market turns.

You become the agent people already trust, before they ever need you.

Coaching Moment

If you're a Secret Agent, you don't need more confidence.

You need a system that keeps you visible without requiring you to become someone you're not.

That's what the next section builds.

In the next section, we'll make this measurable. We'll walk through how clients actually decide who to call and what they look for when they're trying to confirm, "Is this person the safe choice?"

That's where the Trust Audit comes in.

The Trust Audit: How Clients Actually Decide

Most agents imagine the hiring decision as a comparison.

They picture a client weighing credentials, reading reviews, scanning market reports, and thoughtfully selecting the "best" professional.

Sometimes that happens.

More often, the decision looks simpler—and that simplicity is exactly what makes it dangerous for great agents who stay quiet.

In real life, clients usually don't begin with a list of ten agents.

They begin with a moment.

A conversation at dinner. A message from a friend. A sudden job change. A lease renewal notice. A neighbor who put a sign in the yard. A headline about rates. A baby on the way. A divorce. A parent getting older.

Something shifts, and the question becomes real: "Should we move?"

That's when the search starts—not for the best agent in the abstract, but for the safest next step.

The decision happens in two stages

Most hiring decisions happen in two stages:

Stage one: Recall. Who comes to mind?

Stage two: Confirmation. Is that person still the right choice?

The painful part is that many agents focus on confirmation—website polish, testimonials, credentials—without realizing they're losing earlier.

If your name isn't recalled, you never reach the stage where your competence can be evaluated.

This is why visibility isn't optional in a trust economy. It's not about attention. It's about entering the first stage of the decision at all.

How recall is triggered

Recall is rarely triggered by "who is most qualified?"

It's triggered by cues:

- "I've seen them recently."
- "They're active."
- "They talk about my area."
- "They explain things clearly."
- "They feel steady."

Clients don't always articulate it that way, but you can hear it in the language they use:

- "I keep seeing your posts."
- "You seem like you really know what's happening."

- "You explained that thing about appraisals... it made sense."
- "I know you're in the business."

Notice what isn't on that list: awards, designations, and slogans.

Those can help in confirmation, but recall is primarily built by repeated signals that feel useful and trustworthy.

What clients do next: the confirmation loop

Once a client has one or two names in mind, they do something that looks like research but functions more like emotional due diligence.

They're trying to answer one question:

"Will I feel safe with this person?"

They run a confirmation loop that usually includes some combination of:

- Searching your name
- Looking at your website
- Scanning your social media
- Checking reviews
- Noticing whether you look current or stale
- Asking a friend, "Do you know them?"

And here's the key: they aren't just looking for proof that you exist.

They're looking for coherence.

Does everything they see reinforce the same message?

Do you feel consistent across platforms?

Do you feel active?

Do you feel like someone who explains, not someone who advertises?

If the answer is yes, trust increases.

If the answer is no—if your online presence feels generic, outdated, inconsistent, or overly performative—trust doesn't necessarily collapse. It just becomes fragile.

And fragile trust doesn't survive uncertainty.

Stan's story: "It's nuts not to put your phone number down"

Stan has a pet peeve that tells you everything you need to know about how he thinks:

"Sometimes I see agents' websites, and there's no phone number. Just a form. Or it says, 'Send me a message.'"

"That's nuts."

"This is a house. This is probably the biggest financial decision someone is going to make. And you want them to fill out a form and wait for you to get back to them?"

"People want to talk. They want to hear your voice. They want to know you're real. They want to feel like if something goes wrong, they can reach you."

"So I put my phone number everywhere. On my website. In my emails. On my social profiles. I want people to know: if you need me, you can call me. Right now."

That's not a tactic. That's a trust philosophy.

Friction kills trust.

And when the stakes are high, people don't want barriers. They want access.

Field-Tested Principle

"Every extra step between a client and clarity is a chance for trust to leak."

The Trust Audit: What you should be measuring

The easiest way to improve how clients choose you is to stop guessing and start auditing the signals they're actually receiving.

A Trust Audit is not about vanity. It's not "How many followers do I have?"

It's about whether your presence helps a client move from uncertainty to confidence.

Here are the four questions that matter most:

1. Are you findable?

If someone searches your name, do they immediately see that you're active and credible?

Or do they see a half-built profile and an outdated website?

2. Are you current?

Do you look like someone operating in today's market, or like someone who posted six months ago and disappeared?

3. Are you specific?

Does your presence make it obvious what market you serve and what kind of client you understand, or could you be anywhere?

4. Are you useful?

Do you demonstrate judgment—explaining what's happening and what it means—or do you mostly broadcast announcements?

These questions don't require you to be everywhere.

They require you to be coherent.

When those four elements are strong, you become an easy decision. Not because you "sold" yourself, but because you removed friction from the client's confirmation process.

Why coherence beats volume

A common misconception is that trust is a volume game. Post more, say more, show more.

But in real estate, especially in uncertain markets, trust is a coherence game.

Clients don't need you to entertain them. They need you to make sense of what feels confusing.

That's why a smaller amount of consistent, specific, useful communication often outperforms a larger amount of generic content.

It's not about feeding the algorithm. It's about feeding recognition.

The shift that makes everything easier

Once you see the two stages—recall and confirmation—you stop treating marketing like a performance.

You start treating it like a responsibility:

"I want to make it easy for clients to choose the safe option."

And if you're reading this thinking, "Okay, but what do I fix first?"—good. That's the right question.

Because when you run a Trust Audit, you'll usually discover that you don't need to do fifty things.

You need to do three things better:

- Clarify your message
- Increase consistency
- Build a confirmation layer that reinforces competence

That's where the rest of this book goes.

Coaching Moment

Stop asking: "How do I get more attention?"

Start asking: "If someone checks me right now, what do they see—and does it make me feel like the safe choice?"

That's the audit.

Next, we move into the core engine that makes trust compound: expertise.

Not as information, but as teaching—because the agent who can explain is the agent who gets trusted.

END OF PART I

Part I Summary & Transition

We've established the foundation:

1. Trust is no longer assumed — It must be reinforced through presence, not proximity

2. Competence is invisible without signal — Clarity, specificity, and consistency make you legible
3. The Secret Agent problem — Being good in private doesn't help people who can't find you
4. The Trust Audit — Clients decide in two stages: recall, then confirmation

Now we shift from why trust matters to how expertise compounds.

In Part II, we'll show you why teaching is the highest form of trust—and why reducing uncertainty (not just providing service) is what clients are actually paying for.

PART

II

EXPERTISE: THE ONLY TRUST THAT COMPOUNDS

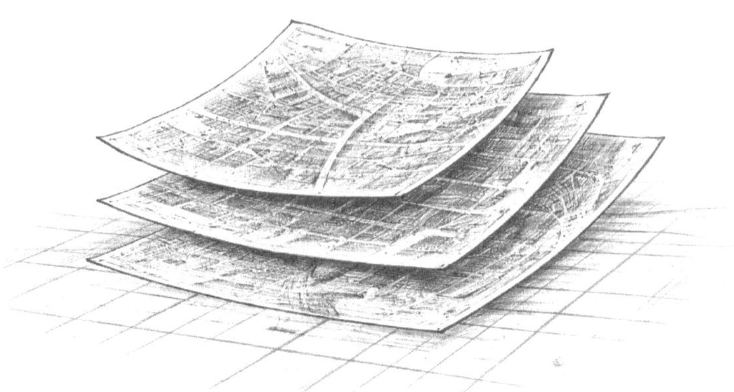

Teaching Is the Highest Form of Trust

A surprising number of clients don't hire an agent because they want someone to "sell their house" or "find them a home."

They hire an agent because they want someone to make the process make sense.

They want someone who can reduce the mental load.

They want to feel like they understand what they're agreeing to—before they sign anything, before they commit emotionally, and before the stakes start to feel real.

This is why teaching is not a nice extra. It's not a branding tactic.

Teaching is the highest form of trust-building available to a real estate professional.

Because when you teach, you're not just giving information. You're demonstrating judgment.

Why information isn't enough anymore

Your clients can access information instantly.

They can search neighborhoods, watch videos, read market commentary, ask a group chat, and get twenty opinions in an afternoon.

The problem is that information is abundant—and interpretation is scarce.

Most of what clients find is either too general to be useful or too extreme to be calming. It's also often disconnected from their actual situation: their finances, their neighborhood, their timing, their risk tolerance, their family, their constraints.

So even after consuming hours of content, they still feel uncertain.

That uncertainty is the real product you're being hired to address.

And the way you address it is by translating complexity into clarity without removing truth.

Stan's story: Pricing explanations that feel fair

Stan has a rule about pricing:

"If I can't explain to the seller why we're pricing at a certain number—and if they can't repeat that explanation back to their spouse—I haven't done my job."

He doesn't just show comps and say, "Here's what it's worth."

He walks them through the why:

- Here's what sold recently in your neighborhood
- Here's how your home compares (size, condition, features)
- Here's what's active right now (your competition)
- Here's what buyers are doing (days on market, offer patterns)
- Here's the tradeoff: price high and wait, or price right and move

And then he asks them a simple question:

"What feels right to you?"

Most of the time, they land on the number he suggested. Not because he pressured them. Because he helped them understand it.

And when the offers come in—or when they don't—there's no conflict. Because the seller already knows why the price made sense.

That's teaching.

And teaching prevents regret.

Field-Tested Principle

"Clarity prevents conflict. Confusion creates chaos."

Teaching creates safety

Think back to the last time you worked with an anxious client.

What helped them wasn't a bigger spreadsheet.

What helped them was when you made the process feel graspable. When you turned the unknown into steps. When you explained a tradeoff plainly. When you took a scary moment—inspection findings, appraisal issues, multiple offer pressure—and gave it context.

Teaching reduces uncertainty in a way that generic reassurance never can.

Saying "Don't worry" doesn't create trust.

Saying "Here's what this means, here's what matters, and here's what we can do next" does.

What clients are really listening for

The clients who trust you most aren't listening for perfection.

They're listening for three things:

1. Do you understand the situation I'm in?
2. Can you explain it in a way I can repeat to my spouse?
3. Do I feel steadier after talking to you than I did before?

If the answer is yes, trust compounds quickly—because you're not just providing service. You're providing clarity.

And clarity is what gets remembered.

The difference between teaching and talking

Many agents "explain" without teaching.

They provide facts. They give details. They answer questions.

But the client walks away with the same feeling they started with: "I'm not sure."

Teaching is different.

Teaching is when you structure the explanation so the client can hold it in their mind. It's when you give them an organizing principle.

For example, instead of listing ten market indicators, you might say:

"What matters right now is not the headline number. It's the relationship between inventory, days on market, and buyer urgency in your segment. If those three are moving in this direction, we price and negotiate one way. If they're moving in the other direction, we adjust."

That kind of explanation doesn't impress people because it's complicated.

It impresses them because it's simplifying something complex into a decision.

That's competence trust.

Stan's story: "Clarity deposits that earn interest"

Stan describes his communication strategy this way:

"Every time I explain something to a client—whether it's pricing, or inspection issues, or how to handle a multiple offer situation—I'm making a deposit."

"Not a sales deposit. A clarity deposit."

"And those deposits earn interest. Because six months later, when they're talking to their friend who's thinking about moving, they don't say, 'Stan was nice.' They say, 'Stan explained it in a way that made sense. He made us feel like we knew what we were doing.'"

"That's the deposit that pays back."

That's teaching as a business strategy.

Not louder. Not more frequent. Just clearer.

Field-Tested Principle

"People don't remember what you said. They remember how clear they felt after you said it."

Expertise that compounds

Here's why teaching is the highest form of trust:

Every clear explanation becomes a deposit that keeps earning interest.

A client might not need you today, but if you helped them understand something that felt confusing, they remember you. They tell someone else. They reference your framing. They come back months later with, "We're ready, and we want you."

This is what it means for expertise to compound.

You're not only closing deals. You're building a reputation for clarity.

And in a trust economy, reputation for clarity is one of the most durable assets you can build.

Coaching Moment

Stop asking: "How do I get more leads?"

Start asking: "How do I become the person who makes hard decisions feel easier?"

That's the whole strategy.

In the next section, we'll go deeper into what your clients are actually buying when they hire you: not just service, but uncertainty reduction—and the specific behaviors that make people feel safe making a decision with you.

Building Rapport Before Expertise Matters

Here's a truth most competence-focused agents resist:

Your expertise is worthless if people don't like you first.

Not because clients are shallow. Because humans make decisions emotionally and justify them rationally.

Before someone trusts your judgment, they have to trust your intentions.

And trust in intentions is built through rapport, not credentials.

The sequence that actually works

Most agents approach new relationships like this:

1. Establish credentials (years in business, sales volume, awards)
2. Demonstrate expertise (market analysis, comps, strategy)
3. Build rapport (if there's time)

That sequence fails more often than it succeeds.

Here's why: when you lead with expertise, you create evaluative distance. The client is assessing you. You're performing competence. The dynamic feels formal, transactional, and slightly adversarial—even if you're both being polite.

The better sequence is this:

1. Build rapport (find commonality, make them comfortable, become human)

2. Establish intentions (show that you care about their outcome, not just your commission)

3. Demonstrate expertise (now they're listening because they trust you)

Rapport disarms the evaluation. It shifts the relationship from "Are you qualified?" to "Can I trust you?"

And once someone trusts your intentions, they're far more willing to trust your expertise.

How rapport actually forms

Rapport isn't manufactured through compliments or small talk.

It's created through moments of genuine human connection—moments where the professional performance drops and you're just two people recognizing something in common.

It might be:

- A shared experience (kids, hobbies, challenges, background)

- A moment of humor (laughter is the fastest rapport-builder)
- Honest vulnerability (admitting something real instead of posturing)
- Recognition (seeing something about them that others miss)

Stan's laugh technique works because laughter is a social signal. It says, "I'm safe. You're safe. We're not adversaries."

Once that signal is sent, the guard drops. The client stops performing "informed buyer" or "savvy seller." You stop performing "expert agent."

And the conversation becomes real.

Why this makes expertise more effective

When you establish rapport first, something shifts in how the client receives your expertise.

Instead of hearing your advice as "sales pitch," they hear it as "guidance from someone who cares."

Instead of questioning your motives, they assume good intent.

Instead of needing to validate every recommendation, they trust that you're steering them correctly.

This doesn't mean they stop thinking. It means they stop defending.

And when defenses are down, teaching becomes possible.

The danger of leading with competence

When you lead with expertise before rapport exists, you risk triggering skepticism.

The client hears your market analysis and thinks, "Are they inflating values to get the listing?"

They hear your pricing strategy and think, "Are they pushing me to sell lower just to close faster?"

They hear your timeline and think, "Are they rushing me because they need the commission?"

None of those suspicions may be true. But in the absence of rapport, clients fill the gap with doubt.

And once doubt enters, every piece of expertise you share has to overcome resistance.

Coaching Moment

Ask yourself: When you meet a potential client, are you trying to impress them—or connect with them?

Impressive gets you evaluated. Connection gets you trusted.

One leads to skepticism. The other leads to partnership.

The AI implication

This is where AI runs into a hard limit.

AI can draft emails. It can generate market reports. It can analyze data and produce content.

But it cannot build rapport.

It cannot read a room and sense when someone needs humor, honesty, or reassurance.

It cannot create the moment when two people lock eyes, laugh at the same thing, and suddenly feel like they're on the same team.

That moment is irreducibly human.

And in an industry increasingly reliant on automation, that moment is your competitive edge.

The agents who win in the AI era won't be the ones who automate everything. They'll be the ones who use AI to handle the repeatable tasks—so they have more energy and presence for the human moments that actually build trust.

Rapport first. Expertise second. AI third.

That's the sequence that compounds.

Reduce Uncertainty, Not Just Provide Service

Most agents describe their value in terms of service: responsiveness, negotiation, follow-up, execution.

Clients appreciate those things, and they matter.

But they aren't the core reason people trust you enough to bet their money, timeline, and emotions on you.

The core product you deliver is simpler and heavier:

You reduce uncertainty.

Real estate isn't stressful because clients don't know what a contract is. It's stressful because they're about to make a decision they can't fully reverse, in a market that changes its mood, with outcomes that don't feel guaranteed.

Even confident people feel exposed when the stakes are high and the variables are moving.

So your true job isn't just to do tasks.

Your job is to turn a high-stakes decision into something that feels navigable.

And the moment you understand that, your communication changes.

Not louder. Not salesier. Not more "content."

Clearer. Calmer. More useful.

Here's the real reason clients panic

Clients rarely panic because of the facts.

They panic because of what the facts might mean.

An inspection report isn't scary because it exists. It's scary because they don't know what's normal, what's expensive, what's a negotiation lever, and what's a red flag.

A slow week of showings isn't scary because it happened. It's scary because they don't know whether it's seasonal, pricing, marketing, or the market turning.

Uncertainty is not a data problem.

It's a meaning problem.

That's why more information doesn't always help. Sometimes it makes things worse.

You can drown a client in market stats and still leave them feeling unstable, because what they needed was interpretation.

They needed someone to say, "Here's what matters. Here's what doesn't. Here's what we do next."

Stan's story: Window shop, top three, door handle

Stan has a process he's refined over twenty-two years. He calls it his filtering system.

When a buyer comes to him, he doesn't start by showing houses.

He starts by asking questions. Fast, precise, narrowing questions:

"One-story or two-story?" "Pool or no pool?" "How many bedrooms?" "What part of town? What school district?" "Open floor plan or traditional?" "How big of a lot?" "How many car garage?"

In five minutes, he's narrowed the entire Las Vegas market from thousands of homes down to twenty.

Then he does something most agents don't:

He sends those twenty listings to the client and says, "Window shop. Give me your top three."

He doesn't make them look at all twenty. He doesn't waste their time. He trusts them to know what resonates.

And then—here's the key—he only shows them three homes.

Why?

Because if he's done his job right, one of those three is going to feel right. And he'll know which one the moment they walk through the door.

"I can see it when they touch the door handle," Stan says. "It's like when you see a puppy and you know. That instant connection. They don't have to say anything. I can feel it."

That's not luck.

That's uncertainty reduction engineered into a process.

Stan didn't give them more options. He gave them clarity.

He removed the fog. He protected their time. He made the decision feel manageable.

And when that happens, clients don't feel pressured. They feel guided.

Field-Tested Principle

"Your job is not to show houses. Your job is to reduce decision fog."

The Uncertainty Ladder

When you're trying to build competence trust, you're guiding people up a ladder:

Chaos → Context → Choices → Commitment

Chaos is, "We don't know what to do."

Context is, "Here's what's normal."

Choices are, "Here are the paths and tradeoffs."

Commitment is, "Okay—let's move."

Most agents stop at context. They explain a few things and assume understanding will create confidence.

But confidence usually arrives at the next rung—choices.

Because choices restore agency.

Clients don't want perfect certainty. They want a sense that they're not trapped. They want to feel like whatever happens, there's a plan.

So your job is to do something subtle but powerful:

Convert stress into decisions.

What pros do in uncertain moments

The most trusted professionals—inside any high-stakes industry—do three things consistently.

You can feel it when you're around them. They don't inflate fear. They don't perform confidence. They don't pretend risk is gone.

They do this:

They name reality. They frame tradeoffs. They define the next step.

That's it.

And when you do that, clients feel held.

Not manipulated. Not rushed. Held.

Here's what that sounds like in real estate language:

- "This inspection item is common. It matters, but it's not fatal. Here's the range we see, and here's how we ask for it."
- "We didn't get the offer accepted. That doesn't mean you 'lost.' It means we learned how this segment is behaving. Here's the adjustment."
- "The market is softer than it was. That doesn't mean we panic. It means we price with precision and negotiate differently. Here's the plan."

Notice: you're not promising outcomes.

You're promising guidance.

That's what trust is.

The difference between being helpful and being stabilizing

Helpful is answering questions.

Stabilizing is organizing the situation.

Helpful is "Here's what the report says."

Stabilizing is "Here's what this means, what's normal, and what we'll do."

Helpful makes clients informed.

Stabilizing makes clients decisive.

And decisive clients don't spiral. They move.

Your brand is the emotional aftertaste

Your brand is not your headshot.

It's the feeling people have after they interact with you.

Do they feel calmer? Do they feel clearer? Do they feel like the decision is manageable?

If yes, congratulations—you're building the rarest brand in real estate:

The agent who makes hard things feel solvable.

That's why some agents can post less and still win more. Their reputation creates gravity. People bring them into conversations because they don't just "know real estate." They reduce uncertainty around it.

And once you own this, your marketing stops feeling like marketing.

Because clarity is a form of service.

Coaching Moment

If you want one change that improves everything:

Stop asking: "How do I get more leads?"

Start asking: "How do I become the person people feel safer with?"

That's the whole game.

Next, we talk about the fastest way to become "safer" in the client's mind before they ever meet you: specificity.

Generalists feel competent. Specialists feel inevitable.

The Expert Position: Why Specific Wins

Most agents are taught to market themselves like generalists, because being broad feels safe.

"I help buyers and sellers." "I serve everyone." "I work all price points."

It sounds inclusive. It sounds professional.

It also makes you forgettable.

Not because you aren't capable.

Because the human brain doesn't store broad claims. It stores categories.

In a trust economy, clients don't want to work hard to decide whether you're a fit. They want the decision to feel obvious. They want to feel like they're choosing the safe option.

Specificity creates that feeling faster than almost anything else.

Specificity is a trust shortcut (and that's okay)

Here's what's happening inside the client's mind:

When they see a generalist, they think, "Maybe."

When they see a specialist, they think, "Probably."

And "probably" is a relief when the stakes are high.

Specificity reduces perceived risk because it implies repetition—experience with the exact situation, the exact client type, the exact neighborhood dynamics, the exact tradeoffs.

It signals: "I've been here before."

That's not a marketing trick. That's how humans choose professionals.

You want the client to feel this sentence without you ever saying it:

"They've handled my problem before I even called."

The Positioning Triangle

A clean expert position usually has three sides:

Where you operate (market / area / segment) Who you help (client type / life stage) What you're known for (the outcome / edge you deliver)

That's it.

You don't need a complicated brand story. You need a clear handle people can remember and repeat.

Because referrals aren't built on your bio.

Referrals are built on someone being able to say, quickly:

"You need to call this person."

Coaching Moment

If your friend can't describe you in one sentence...

They can't refer you under pressure.

And most referrals happen under pressure.

"But I don't want to limit myself..."

You're not limiting your business. You're shaping your reputation.

This is where agents get stuck: they confuse what they do with what they lead with.

You can still take a wide range of transactions. But what you lead with should be what makes you memorable and believable.

Specificity doesn't shrink opportunity—it increases conversion.

When you feel like the obvious choice for someone, you don't just attract those clients. You also become the "safe fallback" for adjacent clients because you feel competent by default.

General claims are easy to say. Specific claims are harder to fake.

That's why specific claims work.

Stan's story: The seized home on the 18th tee

Stan's favorite deal happened because of proximity, trust, and clarity—not marketing.

He'd moved to Las Vegas and joined a country club. He wasn't there to sell. He was there to play golf, meet people, and be part of something.

But he also didn't hide what he did.

One day, he was playing a round with a new acquaintance—someone he'd only met a few times. They walked past a row of three massive homes that had been part of the Street of Dreams showcase a few years earlier.

The homes were now empty. Seized by the federal government. The builder had been involved in some kind of Ponzi scheme. It was a mess.

Stan's golf partner stopped and said, "Man, that house is incredible. I love it."

Stan said, "You want to see it?"

They walked the property. Stan explained what he knew about the situation—the legal complexity, the fact that it was 80% done, the opportunity to finish it custom.

His friend said, "I want to buy it."

Six months of legal work later, they closed. The friend paid $2.8 million for a 10,000-square-foot home that's now worth $15-18 million.

But here's the real story:

That one deal turned into thirty more.

They started flipping homes together. They bought land. They did investment deals. Over the years, Stan estimates he's done $40 million in transaction volume that traces back to that one golf game.

Not because he was the best marketer.

Because he was present, specific, and trusted.

Field-Tested Principle

"Proximity + trust + clarity = outsized outcomes."

Your content should prove judgment, not activity

Once you choose your lanes, the work becomes lighter.

You stop guessing what to post or say, because your role is clear:

You interpret what's happening inside your lanes.

The most powerful content in real estate isn't entertainment. It's not trends. It's not constant posting.

It's translation.

"What happened" and "what it means."

That's what experts do.

They don't just broadcast. They contextualize.

They don't just report stats. They explain implications.

They don't just announce closings. They show lessons.

Local expertise isn't just about loving your neighborhood; it's about analyzing its dynamics. Investor-focused agents, for example, don't just tout a building's charm; they know the rent rolls, cap rates, and future development plans. They can explain how a new transit line might affect values or why one side of a street commands a higher premium. Sharing specific location insights and ROI projections—while interpreting what those numbers mean for your client's goals—positions you as the guide clients want, whether they're buying their first home or acquiring a portfolio.

And that creates competence trust.

The "two-to-three lanes" rule

Most market leaders can be described by two or three lanes.

Not because they refuse other business, but because the market needs a simple story.

Two or three lanes gives you coherence.

Coherence gives you recognition.

Recognition gives you recall.

Recall gives you calls.

This is how a reputation compounds.

The Market Leader Effect

Market leadership isn't always the highest production number.

Often it's something quieter and more durable: being the most trusted interpreter in a defined space.

When uncertainty rises, people don't need more agents.

They need better interpreters.

If you become the interpreter in a lane, you get pulled into decisions before the transaction even exists. Your name shows up in group chats. Your advice gets repeated. Your "way of seeing" becomes a reference point.

That is the Realtor Edge in motion.

Coaching Moment

Your goal is not to be famous.

Your goal is to be the first name that feels safe in a specific situation.

That's the whole strategy.

Next, we move into systems—because insight and specificity don't compound unless you can show up consistently.

And consistency is hard when your life is full and your business is real.

Systems protect your presence. They keep you from disappearing. They make your trust deposits automatic.

END OF PART II

Part II Summary & Transition

We've established how expertise compounds:

1. Teaching is the highest form of trust — Clarity deposits earn interest
2. Reduce uncertainty, not just provide service — Your job is to make decisions feel navigable
3. The expert position: why specific wins — Specificity creates trust shortcuts

Now we shift from what builds trust to how to protect it.

In Part III, we'll build the systems that keep you present even when life gets full—because consistency is what separates agents who compound trust from agents who cycle through chaos.

PART

III

SYSTEMS: THE ARCHITECTURE OF RELEVANCE

The Marketing Equation: Timing + Trust

Most agents approach digital marketing backwards.

They're told to focus on SEO and ranking #1 on Google. They're told to run Facebook ads to generate leads. They're told to capture website traffic and convert strangers into clients.

This is the e-commerce playbook. It works brilliantly for selling shoes online. Someone sees an ad, clicks, buys—transaction complete in 15 minutes.

Real estate doesn't work like this.

The average person buys or sells a home every 7-10 years. When they're "in the market," the decision cycle is 6-18 months. And when they finally decide, they're not choosing the agent with the best SEO or the catchiest Facebook ad.

They're choosing someone they trust.

Think about your last three closings. How did those clients find you?

Most successful agents will say:

- Referral from a past client
- Friend or family connection
- Someone who'd been following their content

- Met at a community event or through mutual connections

Notice what's missing? Nobody says "I ranked #1 for 'homes for sale in Dallas'" or "My Facebook ad was really compelling."

The Lead Generation Trap

Here's what happens when agents chase tactics instead of understanding the game:

Agent A: Spends $2,000/month on Facebook ads targeting active buyers. Thirty "leads" are generated. Three respond (10% response rate). One closes (33% close rate).

Annual investment: $24,000 spent for 6 closings = $4,000 per deal.

Agent B: Takes a different approach. Spends $500/month on an email platform and social tools. Maintains consistent contact with a 400-person sphere. Conservative 4.5% of their sphere refers them annually = 18 referrals/year. Closes 70% of those referrals.

Annual investment: $6,000 spent for 12-13 closings = $500 per deal.

The difference:

- 2x more closings
- 8x lower cost per closing
- Higher client satisfaction
- Compounding returns (happy clients become referral sources)

This doesn't even include the opportunities from community members who discover you through content.

The math isn't close.

So why do agents keep chasing leads instead of building trust?

Because lead generation feels like action. It's immediate. It's measurable. You can point to ad spend and say "I'm marketing."

Trust-building feels slower. The returns are delayed. The attribution is fuzzy.

But here's the reality: Real estate is a timing business wrapped in a trust business.

Why SEO Doesn't Work (For Most Agents)

Let me be direct: For 95% of real estate agents, investing heavily in SEO is wasted money.

When someone searches "homes for sale in Dallas," your competition is:

- Zillow (Domain Authority: 95/100)
- Realtor.com (Domain Authority: 94/100)
- Redfin (Domain Authority: 92/100)
- Your brokerage's national site (Domain Authority: 75/100)

These platforms have:

- Hundreds of millions in investment
- Decades of trust signals with Google
- Millions of backlinks from authoritative sources
- Teams of engineers optimizing every detail

You have a website template and maybe a blog you update quarterly.

You cannot compete on these terms.

The One Exception

Hyperlocal, ultra-specific content can work. If you genuinely become the expert in specific neighborhoods and create comprehensive content consistently, you can rank for:

- "Victorian homes in Oak Cliff Dallas"
- "Living in Lakewood complete guide"
- "Decatur vs Virginia Highland comparison"

But that's not really an SEO strategy. That's a content authority strategy that happens to have SEO benefits.

For 95% of agents: Your time and money deliver better ROI elsewhere.

The Real Game: Timing + Trust

Real estate is fundamentally different from demand generation businesses.

Pizza is demand generation: When someone decides they want pizza, they choose based on convenience, price, or recent advertising. The decision happens in minutes. They'll make the same decision again in a few days.

Real estate is timing and trust: Everyone you know will need a realtor someday. The question isn't "Will they need one?" The question is: "When their moment arrives, will they think of you?"

This changes everything about digital marketing strategy.

You're not trying to convince someone to buy a house today. You're building:

Timing: Top-of-mind awareness so when their moment arrives (6 months, 2 years, 5 years from now), you're who they think of.

Trust: Enough credibility and rapport that choosing you is the obvious decision.

Every digital marketing tactic serves one or both of these objectives.

If it doesn't build timing or trust, it's noise.

The Trust Pyramid: Where Your Effort Actually Pays

Not everyone in your potential audience has the same relationship with you. They exist in tiers, and your digital marketing must serve each differently.

Tier 1: Your Sphere (The Foundation)

Who they are:

- Past clients
- Friends and family
- Professional contacts
- Anyone with personal relationship

What they know:

- You're a realtor
- You're competent
- They like you personally

What they need from you:

- Consistent, non-annoying reminders you exist
- Reasons to think of you when they or someone they know needs help
- Continued evidence you're active and successful

Your goal: When they or their network needs a realtor, you're the only name they think of.

Resource allocation: 50% of your digital marketing effort

Harsh truth: Most agents neglect this tier entirely while chasing strangers online.

Tier 2: Your Community (The Middle)

Who they are:

- People in your target neighborhoods who don't know you
- Members of communities you serve
- People who've discovered your content but haven't engaged

What they need:

- Discovery (finding out you exist)
- Familiarity (seeing you enough to recognize you)
- Credibility (evidence you know your stuff)
- Affinity (reasons to actually like you)

Your goal: Build enough trust that when they enter the market, they reach out to you directly instead of searching "realtors near me."

Resource allocation: 35% of your effort

Tier 3: Active Buyers and Sellers (The Top)

Who they are:

- People actively in market right now
- Searching for immediate help

- Comparing multiple agents

What they need:

- Fast credibility
- Proof of competence
- Easy contact path

Your goal: Convert before they hire someone else.

Resource allocation: 15% of your effort

The mistake: Most agents spend 70% of effort chasing Tier 3 when it's the smallest opportunity.

The Compound Effect: Consistency Beats Intensity

Most agents:

- January: "This year I dominate social!" Posts daily for 3 weeks.
- February: Gets busy. Posts twice.
- March: Feels guilty. Creates 15 posts in one weekend. Burns out.
- April-May: Sporadic posting.
- June: "Social media doesn't work."

Market Leaders:

- Every week, 52 weeks:
    - 3-4 valuable social posts
    - One email to sphere
    - One neighborhood-focused piece of content
    - Daily engagement (15 minutes)

After 12 months:

- 150+ social posts providing value
- 52 emails maintaining top-of-mind
- 52 pieces of neighborhood content establishing authority
- 500+ engagement touchpoints

After 24 months:

- 300+ posts (undeniable presence)
- 104 emails (sphere can't forget you)
- 104 neighborhood pieces (obvious expert)
- 1,000+ touchpoints (relationships at scale)

This is compounding. Each piece builds on the last. Trust accumulates. Authority solidifies.

By year two, you're not convincing anyone you're an expert.

It's simply obvious.

Field-Tested Principle

"Marketing isn't convincing people you're good. Marketing is making it easy for the right people to recognize that you're already good."

Coaching Moment

Stop asking: "How do I get more leads this month?"

Start asking: "How do I become the obvious choice in my community?"

Stop thinking: "What tactics will generate business quickly?"

Start thinking: "What consistent presence compounds over years?"

Stop competing for attention.

Start earning it through value.

In the next section, we'll break down the specific channels—email, social, website, Google—and show you exactly what each one should do in your trust-building system.

The Channel Strategy: Where to Show Up

You understand the equation: Timing + Trust.

You understand the pyramid: Sphere, Community, Active Buyers.

Now the tactical question becomes: Where do you actually show up? And what does each channel do in your trust-building system?

The answer isn't "be everywhere."

The answer is understanding what each channel is built to accomplish—and using it for that specific job.

The Four Pillars of Digital Trust

Trust online is built through four specific mechanisms:

Pillar 1: Consistency (You Show Up)

Posting three times this week then disappearing for a month doesn't feel reliable.

Consistency signals:

- You're still in business
- You're active and successful
- You're committed to this market
- You're not desperate or sporadic

Consistency doesn't mean daily. It means predictable.

Once a week for 52 weeks beats daily for one month then nothing.

Pillar 2: Value (You Help First)

Trust builds through reciprocity. When you help without asking for anything in return, people naturally want to reciprocate.

Every valuable post makes a small deposit in the trust bank:

- Neighborhood guide that helps someone choose where to live
- Market insight that helps timing decisions
- Local business recommendation

Do this 100 times, you've made 100 deposits. When you need a withdrawal (referral), the account is full.

Pillar 3: Credibility (You Know Your Stuff)

People need evidence you're competent.

Credibility comes from:

- Demonstrating deep neighborhood knowledge
- Sharing client success stories
- Explaining complex topics simply
- Showing (not telling) expertise through content

One video walking through a neighborhood pointing out value-adds does more for credibility than 100 posts saying "I'm an expert."

Pillar 4: Humanity (You're Real)

Most agents create such a polished "professional" presence it's sterile. No personality, no vulnerability, no humor.

People don't trust robots. They trust people.

When you share:

- Personal stories about why you love your neighborhood
- Mistakes you made and learned from
- Genuine excitement about client wins
- Your actual personality

You become real. Real people are trustworthy.

Channel Roles: What Each Platform Actually Does

Different channels serve different functions in the trust economy. Understanding their roles prevents wasted effort.

Email = Relationship Preservation + Interpretation Delivery

Purpose:

- Keep sphere warm
- Deliver market interpretation
- Maintain top-of-mind

Why it works:

- You own your list (no algorithm)
- Everyone gets your message (no platform filtering)
- Direct to inbox (high visibility)

Best for: Tier 1 (Sphere) and Tier 2 (Community)

Frequency: Monthly minimum for sphere, more frequent for active leads

What to send:

- Market updates with local interpretation
- Neighborhood spotlights
- Helpful resources (not just listings)
- Personal updates that build connection

What not to send:

- Generic "just checking in" messages
- Automated birthday wishes that feel robotic
- Constant listing promotions
- Anything you wouldn't text a friend

Social Media = Familiarity + Proof of Life + Teaching Snapshots

Purpose:

- Create repeated exposure (familiarity)
- Demonstrate you're active (social proof)
- Provide quick value (teaching moments)

Why it works:

- People scroll daily (repeated touchpoints)
- Visual + personal (builds affinity)
- Low commitment (easy to consume)

Best for: Tier 2 (Community) and Tier 1 (Sphere engagement)

Frequency: 3-5x per week

What to post:

- 70% Value & Community (local spots, weekend activities, personal moments)
- 20% Education & Authority (market insights, Q&A, client stories)
- 10% Direct Business (listings, open houses, calls-to-action)

What not to post:

- Constant "just listed/just sold" spam
- Motivational quotes with stock photos
- Generic content that could belong to any agent
- Overly polished perfection that feels fake

Website = Confirmation + Decision Support

Purpose:

- Confirm trust when people look you up
- Provide depth (comprehensive resources)
- Convert consideration to contact

Why it works:

- People check you before hiring (verification)
- Demonstrates competence (comprehensive content)
- Answers questions (reduces friction)

Best for: Tier 3 (Active) and Tier 2 (Community discovery)

Frequency: Always current, updated monthly minimum

What it needs:

- Clear positioning (who you serve, where, why)
- Recent activity (blog, closings, updates)

- Proof of competence (testimonials, case studies)
- Easy contact (phone number visible, not buried)
- Local specificity (neighborhood knowledge)

What it doesn't need:

- SEO optimization for broad keywords
- Lead capture forms that create friction
- Generic template language
- Bells and whistles that slow load time

Google/Maps = Findability + Legitimacy

Purpose:

- Be discoverable when searched
- Signal legitimacy (reviews, profile)
- Capture local searches

Why it works:

- First place people look (discovery)
- Reviews build trust (social proof)
- Maps shows presence (local authority)

Best for: Tier 3 (Active) and Tier 2 (Community)

Frequency: Maintain accuracy, encourage reviews ongoing

What to optimize:

- Complete Google Business Profile
- Recent photos (monthly updates)
- Active review responses
- Accurate information (hours, phone, website)

- Posts/updates showing activity

What not to do:

- Buy fake reviews
- Ignore negative reviews
- Let profile go stale
- Forget to add your phone number

Metrics That Matter

Leading Indicators (track these):

- Reply rate to your emails
- Branded search volume (people searching your name)
- Returning website visitors
- Call clicks from Google Business Profile
- Direct traffic to website (people typing in your URL)
- Engagement rate on social (engagements ÷ followers)

Vanity Metrics (ignore these):

- Total follower count
- Likes on individual posts
- Random reach spikes
- Impressions without intent
- Traffic from sources that don't convert

Ultimate Metric:

When you meet someone and they say, "I've been following you online!"—that's the real measure of success.

Stan's Philosophy: Touch People the Way They Need to Be Touched

Stan noticed something interesting about his sphere.

Many of his wealthiest, longest-tenured clients—the ones with the most home equity and the highest likelihood of referring business—use AOL and Yahoo email addresses.

What does that tell you?

They're not tech-savvy early adopters. They're probably not on TikTok. They might be on Facebook, but they're not scrolling Reels all day.

Meanwhile, first-time buyers in their late 20s are spending hours on Instagram and TikTok but rarely check email.

The lesson: You have to meet people where they are.

Different demographics consume content differently. Different platforms serve different purposes. And trying to be everywhere dilutes your effectiveness.

Master 2 platforms. Do them well.

Don't try to master five and do them all poorly.

Field-Tested Principle

"Don't chase platforms. Serve your people. When your people move, you move. Until then, double down on what's working."

Coaching Moment

If you're reading this thinking, "I don't have time for all this," you're right.

You don't have time to be everywhere.

But you do have 2-3 hours per month to batch content and 15 minutes per day to stay engaged.

That's not a lot. But done consistently, it's enough to stay top-of-mind with your sphere and build trust with your community.

And in real estate, that's all you need.

Next, we'll build the infrastructure that makes all of this work: your database—the single most valuable asset in your business.

Your Database Is Your Business

Most agents treat their database like a phone book: a list of names they might need someday.

Market leaders treat their database like the most valuable asset in their business. Because it is.

Your database isn't just contact information. It's the compound interest account of your real estate career. Every relationship you add, every interaction you log, every note you capture—it's all equity that appreciates over time.

The problem is that most agents don't manage it like an asset. They manage it like a chore.

This section is about changing that mindset and building a system that turns your database into predictable business.

Stan's Story: From Excel to Systems

When Stan got into real estate in 2004, CRM systems weren't what they are today.

"When I got in the business," Stan says, "doing a CRM was just on an Excel spreadsheet."

He kept track of every client, every contact, every conversation in a simple spreadsheet. Names, phone numbers, addresses, notes about their kids, their timeline, their preferences. Nothing fancy. Just organized.

But here's what mattered: he never let anyone fall through the cracks.

Every past client got a follow-up. Every sphere contact got periodic touchpoints. Every lead got tracked. The system was simple, but it was consistent.

Over time, Stan watched technology evolve. Modern CRM platforms emerged with automation, tagging, workflows, and integrations. He saw agents around him adopt these tools—and he saw just as many agents fail to use them effectively.

The lesson Stan learned: The tool doesn't matter as much as the discipline.

You can have the most expensive CRM in the world, but if you don't use it consistently, it's worthless. And you can have a basic spreadsheet, but if you use it religiously, it becomes the foundation of a multi-million-dollar business.

Today, Stan's perspective has evolved: "You can do a lot more business by using technology." But the technology only works if you understand what you're trying to accomplish.

The goal isn't to "have a CRM." The goal is to never forget anyone, never miss a follow-up, and always know where every relationship stands.

That's what this section is about.

The Database Truth

Here's the reality most agents don't want to hear:

Your database is your business.

Not your website. Not your social media following. Not your personal brand. Not even your listings.

Your database.

Because your database represents:

- Every past client who could refer you
- Every sphere contact who could need you
- Every lead you've ever generated
- Every relationship you've ever built

If you lost your database tomorrow—every name, every note, every interaction history—how long would it take to rebuild your business?

For most agents, the answer is years.

That's why database management isn't a "nice to have." It's the infrastructure your entire business runs on.

Why Most Agents Fail at CRM

Three reasons:

1. They never set it up correctly.

They sign up for a CRM, import contacts in a messy pile, and hope the software magically organizes everything. It doesn't. Garbage in, garbage out.

2. They abandon it when it feels overwhelming.

They start strong, then life gets busy, and the CRM becomes one more thing they "should" be doing but aren't. Guilt builds. They stop logging in.

3. They treat it like a lead generation tool instead of a relationship management tool.

CRMs don't generate leads. They manage relationships. If you're using your CRM like a marketing automation platform and ignoring the relationship layer, you're missing the point.

The agents who win with CRM understand one thing:

A CRM is a memory aid, not a magic wand.

It helps you remember what matters. It helps you follow through. It helps you stay organized when your brain is full.

That's it.

The Five Tiers of Contacts

Not everyone in your database has the same relationship with you. Treating them all the same is inefficient and ineffective.

Here's how to think about your database in tiers:

Tier 1: Past Clients (Your VIPs)

Who they are: Anyone who has closed a transaction with you

Why they matter: Highest likelihood of repeat business and referrals

Communication frequency: 2-4 touches per month

What they need: Continued value, personal connection, proof you're still active

Tier 2: Sphere (Personal Relationships)

Who they are: Friends, family, neighbors, professional contacts

Why they matter: Trust is already built; timing is the only variable

Communication frequency: 1-2 touches per month

What they need: Reminders you're in real estate, occasional value, genuine connection

Tier 3: Active Leads (In Market Now)

Who they are: Buyers/sellers actively working with you

Why they matter: Immediate business opportunity

Communication frequency: Weekly or more, based on need

What they need: Responsiveness, clarity, leadership through the process

Tier 4: Warm Prospects (Future Opportunity)

Who they are: People who've engaged but aren't ready yet (6-24 month timeline)

Why they matter: Will become Tier 3 when timing activates

Communication frequency: 1-2 touches per month

What they need: Education, market updates, continued nurture

Tier 5: Cold Leads (Long-Term Nurture)

Who they are: Old leads, distant contacts, people who showed interest years ago

Why they matter: Low probability individually, but volume creates opportunity

Communication frequency: Monthly at most

What they need: Best content only; don't overwhelm

The mistake most agents make: treating Tier 5 the same as Tier 1.

Your past clients deserve personalized attention. Your cold leads from 2019 do not.

Segment your database. Treat people appropriately based on relationship strength.

The Segmentation Playbook: What to Actually Send

This is where most agents get stuck. They know they should segment, but they don't know what to send to each tier.

Here's your playbook:

TIER 1: PAST CLIENTS

Goal: Referrals + Repeat Business

Communication Frequency: 2-4 touches per month

Type Mix:

- 40% Personal (texts, calls, handwritten notes)
- 40% Value (market updates specific to their neighborhood)
- 20% Social Proof (just sold posts, client wins)

Sample Month:

Week 1: Personal text checking in

- "Hey [Name]! Just thinking about you. How's the house treating you? Any issues with that HVAC we flagged during inspection?"

Week 2: Market update email (specific to their neighborhood)

- "Your neighborhood update: 3 homes sold this month in [Neighborhood], median price up 4% from last quarter. Your home value is estimated at [range]. Let me know if you want a detailed report."

Week 3: Social media tag (if they're okay with it)

- "2 years ago today I helped the [Last Name] family find their dream home in [Neighborhood]. Grateful to be part of your journey! 🏡"

Week 4: "Thinking of you" note with local value

- "Tried the new [Restaurant] on [Street] - thought of you guys! Let me know if you need anything."

Quarterly: Birthday card, home anniversary card, holiday card

Annually: Home valuation report (unsolicited, just FYI)

TIER 2: SPHERE

Goal: Stay Top-of-Mind for When They or Their Network Need Help

Communication Frequency: 1-2 touches per month

Type Mix:

- 60% Value & Community (local insights, helpful info)
- 30% Education (market trends, buyer/seller tips)
- 10% Personal Connection

Sample Month:

Week 1: Social media value post (they see it if following)

- "5 things happening in [City] this weekend" + tag relevant people

Week 3: Monthly email with market snapshot

- "Quick market update for [City]: Here's what's happening and what it means for homeowners"

Quarterly: Personal check-in text or call

- "Hey! Been a while. How are things? Still loving the house?"

Note: Don't ask for business. Just stay present and useful.

TIER 3: ACTIVE LEADS

Goal: Close the Transaction, Deliver Exceptional Experience

Communication Frequency: Weekly minimum, daily during critical phases

Type Mix:

- 80% Transactional Updates (where we are in process, next steps)
- 20% Relationship Building (learning about them, connecting personally)

Sample Month (Pre-Closing):

Daily/Every Other Day: Text or call updates

- "Just got the inspection report. Calling you in 5 minutes to walk through it."
- "Appraisal came in at list price. We're clear to close on schedule."

Weekly: Email recap

- "Here's where we are, here's what's next, here's what you need to do."

Post-Showing: Immediate follow-up

- "What did you think? Do you want to see it again or keep looking?"

Post-Closing: Transition to Tier 1

- Move immediately to past client nurture sequence

TIER 4: WARM PROSPECTS

Goal: Stay Top-of-Mind Until Timing Activates

Communication Frequency: 1-2 touches per month

Type Mix:

- 80% Education (neighborhood guides, market insights, process education)
- 20% Nurture (personal connection, no pressure)

Sample Month:

Week 1: Neighborhood spotlight email

- "You mentioned you were interested in [Neighborhood]. Here's what's happening there right now: inventory trends, new listings, price changes."

Week 3: Educational social media post (they see it if following)

- "Common first-time buyer question: How much do I really need for a down payment?"

Quarterly: Personal check-in

- "Hey [Name], just wanted to check in. Still planning to move in the next 6-12 months? Anything I can help with?"

Note: Never pressure. Just stay useful and present.

TIER 5: COLD LEADS

Goal: Low-Effort Nurture, Hope for Re-Engagement

Communication Frequency: Monthly maximum

Type Mix:

- 100% Best Content Only (your absolute best market insights, guides, resources)

Sample Month:

Once per month: Email with your best content

- "Your monthly [City] real estate update: 3 things every homeowner should know right now"

Social media: They see your posts if they follow (no direct outreach)

Quarterly: Review and prune

- Delete unengaged contacts who haven't opened in 12+ months
- Move engaged contacts to Tier 4

Note: Don't waste energy here. Focus on Tiers 1-3.

The Follow-Up System: Nothing Falls Through Cracks

This is where most agents lose money: they forget to follow up.

A lead comes in. You respond once. They don't reply. You assume they're not interested. You move on.

Reality: They got busy. They're not ready yet. They're still researching. Or they simply forgot to respond.

If you don't have a follow-up system, you're leaving money on the table.

Here's a basic follow-up sequence for new leads:

Day 1 (immediate): Respond within 5 minutes if possible, within 24 hours maximum

Day 2: Follow-up call or text: "Just checking in—did you have any questions?"

Day 4: Value-add: Send helpful resource (neighborhood guide, market update, article)

Day 7: Check-in: "Where are you in your process? How can I help?"

Day 14: Soft CTA: "Want to hop on a quick call to discuss your timeline?"

Day 30: Transition to warm prospect nurture (monthly touches)

The key: automate the reminders, personalize the message.

Your CRM should remind you to follow up. You should write the actual message like a human.

Stan's Approach: "I Schedule It Like a Closing"

Stan treats his marketing system the way he treats a closing: non-negotiable.

"I have a recurring block on my calendar every Friday: 'Marketing Hour.' It's sacred. I don't book appointments during it. I don't let other things creep in."

"During that hour, I review my database, I send follow-ups, I check my content calendar, I batch a few posts for next week. That's it."

"One hour per week keeps my presence alive. And when I get busy, that one hour is what prevents me from disappearing."

That discipline—treating visibility as infrastructure, not inspiration—is what separates agents who compound trust from agents who cycle through chaos.

Field-Tested Principle

"Consistency isn't about intensity. It's about having a floor you never drop below, no matter how busy you get."

Coaching Moment

Your database is the asset you own. Everything else—social followers, website traffic, platform algorithms—can disappear overnight. Your database is yours forever. Treat it like the foundation of your business, because it is.

Next, we'll build the content and communication rhythms that keep your database warm—without requiring you to live online or become someone you're not.

Consistency Beats Intensity

The biggest lie in real estate marketing is that you need to post daily.

You don't.

What you need is a system that keeps you present at a minimum viable level—consistently—without burning you out.

This section is about building that system.

The Busy-Season Disappearance Problem

Here's the cycle most agents live:

Q1: Slow season. Lots of time. You commit to "crushing social media this year." You post daily for three weeks. You feel productive.

Q2: Business picks up. You're showing homes, writing offers, managing inspections. Posting drops to twice a week, then once, then not at all.

Q3: You're slammed. Four closings happening simultaneously. Your social media goes dark. Your email list hasn't heard from you in six weeks.

Q4: Things slow down. You realize your pipeline is thin. You panic. You create 20 posts in one weekend. You burn out. You promise to do better next year.

This pattern doesn't just hurt your visibility. It erodes trust.

.Because inconsistency is interpreted as unreliability.

When your sphere sees sporadic activity followed by silence, they don't think, "They must be busy closing deals."

They think, "Are they still in business?"

That doubt is expensive.

The Minimum Viable Presence Standard

Here's the truth: you don't need to be everywhere, all the time.

You need to be somewhere, consistently.

Minimum viable presence looks like this:

Social Media: 3 posts per week (not daily)

Email: 1-2 emails per month to your sphere

Engagement: 15 minutes daily (responding, liking, commenting)

Content Creation: 2-3 hours per month (batched)

That's it.

Notice what's missing: daily posting, constant Stories, being "always on."

You're not trying to be an influencer. You're trying to stay top-of-mind with people who already know you.

That's a different game. And it requires far less intensity than you think.

The Batching System

The secret to consistency isn't motivation. It's batching.

Most agents approach content creation reactively:

Monday morning: "I should probably post something today."

They scroll for inspiration. They stare at a blank screen. They write something mediocre. They feel guilty about not doing more.

That's exhausting.

Market leaders approach content systematically:

Once per month (90 minutes): Plan the month's content. Decide topics, themes, and key messages.

Once per week (2-3 hours): Create content in batches. Write 3-4 captions. Film 2-3 videos. Schedule everything.

Daily (15 minutes): Post what's already created. Engage with your community.

The batching advantage: You're in "creation mode" once, not scattered throughout the week.

You film three videos back-to-back instead of setting up equipment three separate times.

You write three captions in one sitting instead of staring at a blank screen three different days.

Batching collapses effort and eliminates decision fatigue.

The 70-20-10 Content Framework

You don't need 47 types of content. You need three categories that serve different purposes:

70% — Value & Community

Content that has nothing to do with buying or selling homes, but everything to do with why people should know, like, and trust you.

Examples:

- Local restaurant recommendations
- Weekend activity guides
- Hidden gems in your city
- Community news
- Personal stories
- Neighborhood facts

Mindset: "How can I be genuinely useful to my community, regardless of whether they ever hire me?"

20% — Education & Authority

Demonstrate competence without being salesy.

Examples:

- Market insights explained simply
- Client testimonials (video preferred)
- Educational tips for buyers/sellers
- Problem-solving stories
- Q&A responses

10% — Direct Business

Your listings, open houses, calls-to-action.

Examples:

- New listings
- Just sold celebrations
- Open house announcements

Why 10% works: If you provide massive value 90% of the time, people don't mind (and even appreciate) occasional listing posts.

Stan's Philosophy: Touch People the Way They Need to Be Touched

Stan noticed something interesting about his sphere.

Many of his wealthiest, longest-tenured clients—the ones with the most home equity and the highest likelihood of referring business—use AOL and Yahoo email addresses.

What does that tell you?

They're not tech-savvy early adopters. They're probably not on TikTok. They might be on Facebook, but they're not scrolling Reels all day.

Meanwhile, first-time buyers in their late 20s are spending hours on Instagram and TikTok but rarely check email.

The lesson: You have to meet people where they are.

Different demographics consume content differently. Different platforms serve different purposes. And trying to be everywhere dilutes your effectiveness.

Master 2 platforms. Do them well.

Don't try to master five and do them all poorly.

Field-Tested Principle

"Consistency isn't about doing everything. It's about doing the right things repeatedly."

Coaching Moment

If you're reading this thinking, "I don't have time for all this," you're right.

You don't have time to be everywhere.

But you do have 2-3 hours per month to batch content and 15 minutes per day to stay engaged.

That's not a lot. But done consistently, it's enough to stay top-of-mind with your sphere and build trust with your community.

And in real estate, that's all you need.

END OF PART III

Part III Summary & Transition

We've built the systems that protect consistency:

1. The Marketing Equation: Timing + Trust — Understanding the real game of real estate marketing
2. Your Database Is Your Business — The infrastructure everything else runs on
3. Consistency Beats Intensity — Minimum viable presence that compounds over time

Now we shift from systems to the tool that can protect those systems: AI.

In Part IV, we'll position AI correctly—as an engine that protects your consistency without flattening your identity, if you use it with judgment.

PART

IV

AI: THE ENGINE WITHOUT THE EGO

AI Does Not Create Trust (But It Scales What You Are)

Most real estate AI conversations are framed like this:

"What tool should I use?" "What prompt gets the best results?" "How do I automate my marketing?"

Those are understandable questions, but they're not first questions. They're second questions.

Because if you start with tools, you'll accidentally treat AI like a substitute for judgment instead of what it really is: a multiplier of whatever you're already bringing to the table.

AI doesn't create trust. It can't.

Trust is a human conclusion formed from consistency, clarity, and credibility over time.

AI can't live in your market. It can't feel the difference between a "technically correct" message and a message that lands with your people. It can't sense when a client is anxious and needs reassurance versus when they're avoidant and need structure.

Those things are not data problems. They're human problems.

So the right way to approach AI is not as a content machine.

It's as an engine that protects the thing the market actually rewards: consistent, high-quality communication—delivered in your voice, rooted in your standards, aligned with your positioning.

That's the promise.

And that's also the danger.

Because an engine makes you faster, but it doesn't make you wise.

Why AI feels like magic (and why it isn't)

AI feels like magic because it collapses effort.

Things that used to take an hour can take five minutes. The blank page disappears. The first draft shows up. The brainstorming happens instantly.

You can generate variations, titles, scripts, captions, emails, landing pages, and outlines on demand.

That's real.

But here's what people learn the hard way: speed is not the same thing as quality, and output is not the same thing as signal.

If the input is generic, the output will be generic. If the instructions are vague, the result will be vague. If your positioning is unclear, AI will give you content that could belong to anyone. If your standards are not explicit, AI will happily produce something that sounds plausible while subtly undermining your credibility.

AI will not protect your reputation.

You must.

The AI mirror effect

Here's the simplest mental model for AI in this business:

AI is a mirror with a megaphone.

It reflects back what you give it—your thinking, your clarity, your taste, your structure, your assumptions—and then it projects it at scale.

If you're thoughtful, it can make you prolific without diluting you. If you're sloppy, it can make you loud without making you credible.

That's why AI can be an advantage for serious agents and a liability for everyone else.

And that's why the first step isn't "what tool."

The first step is deciding what you want scaled.

What should be scaled in a trust economy

In a trust economy, there are only a few things worth scaling:

Clarity. The ability to explain what's happening and what it means.

Consistency. The ability to show up steadily without burning out.

Specificity. The ability to sound like you live in your lane—not like a template.

Standards. The ability to communicate in a way that protects credibility.

AI can help you scale all four, but only if you set the rules.

Which brings us to the real dividing line between agents who benefit from AI and agents who get harmed by it:

The agents who win with AI treat it like an assistant that needs training. The agents who lose treat it like an intern they never supervise.

Three traps that make AI backfire

Trap One: Tool-Chasing

Agents hop from tool to tool, trying to find the magic app that "fixes marketing."

They end up with scattered output and no coherent signal.

The market doesn't reward variety. It rewards trust.

And trust is built through repetition of a clear message over time.

Trap Two: Outsourcing Identity

If AI is writing in a voice that doesn't sound like you, your audience feels it.

They may not say it out loud, but they register the mismatch.

Trust isn't only built by what you say—it's built by the sense that it's coming from a real person with consistent thinking.

Trap Three: Confusing Output with Presence

AI can help you produce a lot.

But producing a lot doesn't mean you're building trust.

Trust is built when your output is useful, coherent, and consistent long enough for your market to associate you with clarity.

AI can make you visible.

Only you can make you credible.

Field-Tested Principle

"AI doesn't replace judgment. It protects continuity."

Coaching Moment

If AI made you twice as visible tomorrow...

Would your market trust you more—or would they just see more of you?

That question is the difference between using AI as leverage and using AI as noise.

AI vs. Google: The most important distinction

This is where many agents misunderstand what AI is doing.

They treat it like an information source, like a smarter search engine.

In reality, AI is best used as a communication engine.

Google finds information. AI helps you transform information into explanation.

That's why AI is useful for real estate professionals.

Your value is not "having data." Your value is interpreting data in context and communicating it in a way that reduces uncertainty.

AI can help you draft that communication faster—emails, scripts, market notes, social posts, FAQs, client updates—without lowering your standards.

But only if you're directing it toward explanation, not just output.

Stan's story: "I still check every number"

Stan uses AI to help him draft market updates.

He feeds it data—recent sales, inventory trends, days on market, interest rate movement—and asks it to create a first draft of an email to his sphere.

But he doesn't send it as-is.

"I always check the numbers. Always. Because AI will confidently tell you something that's wrong."

"It'll say inventory is up 12% when it's actually up 8%. Or it'll say the median price dropped when it was flat. Small mistakes. But in real estate, small mistakes kill credibility."

"So I use AI to get the structure. To get the flow. To get a clean first draft. But I verify every claim. I add the local context. I make sure it sounds like me."

"Because at the end of the day, my name is on it. And my clients trust me because I tell them the truth, not because I post fast."

That's the right use of AI.

Speed without sloppiness. Leverage without laziness.

Field-Tested Principle

"AI writes the draft. You write the credibility."

The hallucination rule

AI will sometimes sound confident and be wrong.

This is called a hallucination—when the model generates something plausible but factually incorrect.

In real estate, hallucinations are dangerous because they often involve:

- Local facts (inventory, DOM, rate commentary, neighborhood details)
- Claims ("top agent," "#1," "best," awards)
- Statistics and sources
- Anything that could create liability

Before you publish or send AI-assisted content, verify:

- Every number
- Every claim
- Every statistic
- Every recommendation

If you wouldn't stake your reputation on it, don't let AI send it on your behalf.

What never gets automated

Here's the bright line:

AI can assist. You decide.

If you wouldn't delegate a task to an unlicensed assistant, don't delegate it to a model.

Never automate:

- Pricing recommendations
- Legal interpretations or contract language
- Fair housing risk areas (steering language, "good schools," "safe neighborhoods")
- Negotiation strategy without oversight

- Disclosure-sensitive advice
- Client confidential data

AI can help you organize thoughts, draft explanations, and structure communication.

But final decisions—especially those involving price, liability, compliance, or client trust—must always route through human judgment.

The trust protection checklist (quick)

Before anything goes live, ask:

- Does this sound like me—or like a template?
- Is it specific to my market and lane?
- Does it reduce uncertainty or increase anxiety?
- Is every factual claim verifiable?
- Would I be comfortable reading this out loud to a client?
- Would my broker/compliance officer approve it?
- Did I remove any sensitive client info?

If you can't confidently answer yes, don't publish yet.

Coaching Moment

Don't ask AI to "write my content."

Ask it to "help me say this clearly, in my voice, for my people."

That subtle shift protects your credibility.

The goal of AI in this book

The goal is not to automate your business into something impersonal.

The goal is to protect what makes you trusted:

- Your voice
- Your clarity
- Your consistency
- Your standards
- Your ability to reduce uncertainty

If AI can help you keep showing up when you're busy, when you're tired, when the market is noisy—then it becomes a real advantage.

If it makes you sound generic, it becomes a reputation risk.

Next, we'll focus on the best use of AI for serious agents: not "content creation," but consistency protection.

AI as the system that keeps your trust deposits happening even when life gets full.

AI as a Consistency Engine

If you take nothing else from this section, take this:

The best use of AI in real estate is not creativity.

It's continuity.

Most agents don't disappear because they don't care.

They disappear because their business gets real. Closings stack. Clients need attention. Family needs attention. Your brain gets saturated.

The first thing to go is visibility, because visibility feels optional compared to the urgent needs of active deals.

But your market doesn't experience that as "they're busy."

It experiences it as silence.

And silence causes trust to cool.

This is the central problem AI can solve ethically: it can reduce the friction between your intent to show up and your ability to show up.

AI doesn't replace your thinking. It supports your execution.

The "first draft" advantage

The most valuable thing AI can do for a working agent is produce a decent first draft that you can refine quickly.

That sounds small until you've lived through the reality of trying to write a thoughtful market update after a ten-hour day.

The blank page is expensive. It costs energy you don't have.

AI removes the blank page.

It gives you something to react to—something you can shape, correct, and bring into your voice.

It turns "I should write that" into "Let me fix this and send it."

That is not automation. That is leverage.

What consistency actually needs

Consistency doesn't need perfection. It needs a repeatable process.

If you build your visibility around "every post must be brilliant," you'll disappear.

If you build it around "I can produce a clear, useful draft quickly, and I'll improve it with judgment," you'll stay present.

That's what AI enables: a stable baseline.

You still provide the standards. You still provide the truth. You still provide the market nuance.

AI provides momentum.

Stan's story: "AI helps me stay consistent when I'm slammed"

Stan uses AI to keep his email cadence alive even during his busiest weeks.

"I'll have three closings happening at once, inspections, appraisals, negotiations—and I know I should send my weekly market note, but I just don't have two hours to sit down and write."

"So I use AI. I give it the data. I tell it what I want to say. It gives me a draft in five minutes."

"Then I fix it. I add the local detail. I tighten the language. I make sure it sounds like me. I verify the numbers."

"And I send it."

"Without AI, I'd skip the email. I'd tell myself I'll do it next week. And next week I'd be slammed again. And before I know it, I've disappeared for a month."

"AI doesn't make me sound smarter. It makes me stay present."

That's the use case.

Not replacing Stan. Protecting Stan's presence when his attention is needed elsewhere.

Field-Tested Principle

"AI doesn't make you louder. It keeps you from going silent."

The four things AI should help you do (and the one thing it shouldn't)

AI is best used to support:

1. Turning market facts into client-friendly explanations
2. Turning your opinions into structured communication
3. Turning scattered thoughts into clear teaching moments
4. Turning one idea into multiple formats (email, post, script) without reinventing it

What it should not do:

Replace your judgment—especially when you're making decisions that affect price, negotiation, legal obligations, or fair housing language.

AI can assist. You decide.

The "voice drift" problem—and how to prevent it

If you use AI casually, you'll notice something:

Your content starts to sound... like AI.

Even if it's "good," it loses the texture that makes people trust you. It becomes smooth, generic, and strangely unmemorable.

That drift is not inevitable.

It happens because most agents never define their voice, their standards, or their positioning clearly enough for AI to follow.

The fix is training, not prompting.

That's the next section.

But even before training, there's a simple discipline that keeps you safe:

AI writes the draft. You write the credibility.

You add the local detail. You add the real story. You add the edge cases. You add the line that only you would say.

That's how you keep the output human.

A practical example (without making this a manual)

A client asks: "Should we wait for rates to come down?"

You can answer that a dozen ways, and half of them will either scare them or feel evasive.

A strong answer requires context, tradeoffs, and a next step.

AI can draft the structure quickly, but you provide the truth of your market and the tone that calms rather than sells.

You're not trying to be clever.

You're trying to be stabilizing.

And when you show up that way repeatedly, you stop being one of many agents.

You become the one who makes things make sense.

Coaching Moment

AI is not here to make you sound smarter.

It's here to make it easier to stay present when life gets full.

That's the edge.

Next, we'll build the highest-leverage skill in the AI era: moving from random prompting to a trained assistant—so AI can support your consistency without diluting your identity.

From Tools to Trained Assistant

Most agents who try AI for the first time have the same experience:

It's impressive. It's fast. And it's... not quite right.

It sounds a little too polished. A little too generic. A little too "professional" in a way that doesn't match how they actually talk.

Or worse, it confidently writes something that is technically plausible but subtly misaligned with the way they would advise a real client.

That gap is the whole game.

It's the difference between AI being a novelty and AI becoming infrastructure.

Because the real leverage isn't in asking AI to create something from nothing.

The real leverage is in having AI consistently support the way you think—without drifting, without flattening your voice, and without inventing certainty where you would normally be careful.

That's what a trained assistant is.

Prompting is transactional. Training is relational.

Prompting is what you do when you ask AI a question and hope it guesses what you meant.

You describe the task each time. You correct it each time. You rewrite it each time.

It works, but it's tiring. It's like working with a brand-new assistant every morning.

Training is different.

Training is when you give the system enough context that it can predict your preferences, honor your boundaries, and produce drafts that feel like you—not like a template.

Here's the simplest way to say it:

Prompting creates output. Training protects identity.

And identity is the trust layer.

The "Context Packet" (the thing most people skip)

If you want AI to help you without diluting you, you need a single source of truth that tells it who it's working for.

We call that a Context Packet.

It isn't complicated. It's just explicit.

A strong Context Packet includes four categories:

1) Voice

How you sound when you're at your best: calm, direct, clear, not hypey.

What you avoid: buzzwords, clichés, dramatic urgency, performative confidence.

2) Positioning

Your lanes. Your market. Your client type.

The kinds of transactions you understand deeply. The kinds you avoid.

The things you are known for—clarity, negotiation strategy, process leadership, calm guidance.

3) Philosophy

How you advise people. How you frame tradeoffs. How you handle uncertainty.

What you believe about timing, pricing, risk, and client psychology.

4) Boundaries

What AI should never decide alone. What needs verification. What language to avoid. How you stay compliant. Where you refuse to oversimplify.

That last piece matters more than people realize.

Clients trust you not because you always sound confident. They trust you because you know where confidence ends and care begins.

AI needs those edges.

Why generic AI output is a reputation risk

Generic content doesn't just fail to impress.

It can actively weaken trust because it signals a lack of specificity.

And specificity is one of the main reasons people feel safe hiring you.

If your website, emails, or posts start to sound like they could have been written by anyone, you lose the thing you're trying to build: the sense that you are a real professional with real judgment in a real market.

This is why "more content" isn't the answer.

More generic content simply scales sameness.

The goal is fewer, better signals—produced consistently.

A trained assistant makes that sustainable.

Training AI to sound like you (without turning into a tech project)

Training doesn't mean you need to become technical.

It means you need to become clear.

You don't need fifty prompts. You need a few strong examples and a few strong rules.

The fastest way to train an assistant is to feed it "gold standard" samples—emails you're proud of, explanations that landed well, a market note that clients responded to—and then define what made them effective.

You're essentially saying, "This is what I mean by clarity. This is what I mean by calm. This is what I mean by useful."

The second fastest way is to give it the negative examples too.

Not the full embarrassing drafts, but the patterns you refuse to use:

- No "crushing it" language
- No manufactured urgency

- No empty motivational fluff
- No fake certainty
- No generic "buy now" advice
- No one-size-fits-all market commentary

You're setting guardrails.

And guardrails are the difference between AI as leverage and AI as liability.

Stan's story: "I taught it how I talk about pricing"

Stan built a simple Context Packet for pricing conversations.

He gave AI three examples of how he explains pricing to sellers—complete with his tone, his structure, and his philosophy.

Then he gave it three rules:

1. Always explain the "why," not just the number
2. Frame it as a tradeoff, not a command
3. End with a question, not a demand

Now, when Stan needs to draft a pricing email or a consultation script, he asks AI to follow his framework.

The result?

"It sounds like me. Not perfectly. But close enough that I only have to tweak a few sentences instead of starting from scratch."

"And that saves me an hour every week."

That's training.

Not magic. Not automation.

Leverage.

Field-Tested Principle

"AI doesn't need to be perfect. It needs to be close enough that you can refine it quickly."

The moment this becomes real

When your assistant is trained, something changes in your weekly life.

You stop staring at a blank page.

Instead, you work like a professional editor:

- You dictate a few bullet points
- You generate a draft
- You correct the local nuance
- You tighten the language
- You send

That sounds simple. It is.

And it's powerful because it protects consistency—the thing we've already established as the trust asset.

This is how AI supports the Realtor Edge: it keeps your signals alive when your schedule would otherwise erase them.

The ethical edge: clarity without manipulation

There's also an ethical dimension to all of this.

The point of systems and AI is not to "manufacture" trust.

It's to make real competence easier to recognize.

When you're training your assistant, you're not teaching it to persuade people into a decision.

You're teaching it to help you communicate like a responsible professional:

- Accurately
- Calmly
- With context
- With tradeoffs
- With next steps
- Without hype

That kind of communication doesn't manipulate. It stabilizes.

And stabilization is exactly what clients are paying for.

Coaching Moment

If you're going to use AI, you have to choose:

Do you want it to make you faster... or do you want it to make you more you at scale?

The agents who win choose the second option.

The bridge to what comes next

At this point in the book, you should feel a clear shift.

We started with trust. We moved into expertise. We built systems to keep trust from decaying. Then we introduced AI as the engine that makes consistency easier without flattening identity.

Now we're going to move into the place where most agents unintentionally lose trust: visibility.

Because visibility is unavoidable. Clients will look you up. They will scan your presence. They will decide what to believe before they ever speak with you.

The goal isn't to be everywhere.

The goal is to be legible—so when people check, what they see confirms what your best clients already know.

That's where we go next.

END OF PART IV

Part IV Summary & Transition

We've positioned AI correctly:

1. AI Does Not Create Trust — It's a mirror with a megaphone; it scales what you already are
2. AI as a Consistency Engine — The best use is continuity, not creativity
3. From Tools to Trained Assistant — Training protects identity; prompting is transactional

Now we shift from how AI supports systems to how to make trust visible.

In Part V, we'll show you how to make your competence legible—so when clients check (and they will), what they see confirms that you're the safe choice.

PART

V

MAKING TRUST LEGIBLE

Visibility Without Credibility Is a Liability

Most agents think the visibility problem is solved by being seen more.

Post more. Shoot more video. Run ads. Be consistent.

But visibility on its own doesn't create trust.

In some cases, it does the opposite. It creates exposure without credibility, and exposure without credibility is a liability.

The modern client doesn't just see you. They evaluate you.

They're not asking, "Does this agent exist?"

They're asking, "Is this agent safe?"

And "safe" doesn't mean perfect—it means coherent, current, and credible.

This is why the worst kind of visibility is visibility that feels empty. It may create familiarity, but it doesn't create competence trust.

And familiarity without competence trust is fragile. It evaporates at the first sign of uncertainty.

The quiet evaluation

Most agents underestimate how clients evaluate them online because the process is quiet.

It happens in a few seconds, often without a message, without a like, without any visible proof it happened at all.

A client hears your name, gets curious, and does what modern humans do: they check.

They Google you. They click your website. They glance at your social profiles. They scan reviews. They look for signals that answer one question:

"Is this person safe to trust with a high-stakes decision?"

And what they're looking for is not perfection.

They're looking for coherence.

Coherence is when your presence tells a consistent story: you're active, you're current, you're specific, and you communicate like a professional who can reduce uncertainty.

When coherence is present, clients feel a subtle relief. They may not say it out loud, but they think, "Okay. This makes sense."

When coherence is missing, they don't always reject you—they just hesitate.

And hesitation kills decisions.

This is why visibility without credibility is a liability. It creates exposure that can actually increase doubt.

The difference between being seen and being trusted

Being seen is easy. Trust is harder.

You can be seen through volume: more posts, more ads, more video, more content.

But if the content doesn't demonstrate judgment, volume simply accelerates sameness.

It can make you familiar without making you credible, and that's a fragile position to be in—because familiarity can be replaced overnight by the next loudest thing.

Credibility isn't created by being interesting. It's created by being useful in a very specific way:

You reduce uncertainty. You help people understand what matters. You frame tradeoffs. You show standards. You feel steady.

That steadiness is what people remember when the timing becomes real.

Stan's story: Physical visibility as memory reinforcement

Stan doesn't just rely on digital presence.

He uses physical visibility too—not as ego, but as memory reinforcement.

"I have billboards. I have a wrapped vehicle. I have signs in yards. And people ask me, 'Stan, do those really work?'"

"And I tell them: it's not about the billboard getting me a call tomorrow. It's about the billboard making sure people remember I'm active."

"When someone sees my face on a billboard on their way to work, and then they see my post on social media, and then they get my email—it all adds up."

"They're not necessarily thinking about moving today. But when they are, they don't have to work hard to remember me. I'm already there."

That's visibility done right.

Not loud. Not desperate. Just present.

He also shares a specific ROI story that proves this works:

He invested $6,000 in billboard placement. That investment resulted in $400,000 in commission income—deals that came directly from people who said, "I kept seeing your billboard."

That's a 66x return.

But here's what most agents miss when they hear that story: the bil board didn't work in isolation.

The billboard got attention. Stan's credibility closed the deal.

The lesson: Visibility gets you noticed. Credibility gets you hired.

Field-Tested Principle

"Visibility isn't about being everywhere. It's about being coherent enough that when people check, trust increases instead of leaking."

The "trust confirmation stack"

Think of your online presence as a stack.

Clients don't evaluate you based on one platform. They build a composite impression from whatever they happen to touch first.

For most people, the stack looks like this:

- Search results (Google and Maps)

- Your website
- Your social profiles (even if they don't follow you)
- Reviews or testimonials
- Signs of recent activity

Here's the important part: the client isn't reading deeply. They're scanning for risk. They're looking for red flags, not literary brilliance.

So your job is not to impress them with complexity.

Your job is to remove friction from their confirmation process.

You want their internal conversation to sound like this:

"Okay—this person is real." "Okay—they're active." "Okay—they know this market." "Okay—they explain things clearly." "Okay—I can see what it would feel like to work with them."

That sequence is the sale.

Not pressure. Not persuasion. Confirmation.

Where credibility leaks

Credibility leaks are usually small, but they compound.

They're not dramatic errors. They're subtle signs of neglect or generic positioning that create a question mark.

Here's what a credibility leak looks like in the wild:

You look inactive. Your last social post was four months ago. Your website says "2023 Market Trends." Your Google Business Profile hasn't been updated in a year.

You look vague. Your bio could describe any agent in any city. Nothing specific about neighborhoods, client types, or your approach.

You look like you could be anywhere. No local references. No neighborhood knowledge. No proof you actually live and work in this market.

You look like you're posting to post. Generic motivational quotes. Stock photos. Content that feels mass-produced.

You look like you're performing instead of guiding. Overly salesy language. "I'm the BEST!" claims without proof. Hype without substance.

And again, the point is not to be "a brand."

The point is to communicate like someone who is safe.

The Minimum Viable Credibility standard

This book isn't asking you to become an influencer.

It's asking you to meet a minimum standard of professional legibility.

That standard is surprisingly achievable if you stop chasing tactics and focus on what the client actually needs to see.

Minimum viable credibility looks like:

Clear positioning: A statement of who you serve and where. Not clever. Just clear.

Evidence of activity: Proof you're current. Recent posts, recent closings, recent content.

Demonstration of judgment: Language that shows you think clearly, explain well, and guide calmly.

Proof of standards: Reviews, outcomes, testimonials that show people trust you.

Maintained presence: A digital footprint that doesn't feel abandoned.

When you hit that standard, something powerful happens:

You don't have to convince people you're credible. Your presence does it quietly, before you ever speak.

And once credibility is handled, visibility becomes an advantage instead of a risk.

Coaching Moment

If a stranger looked you up tonight—right now—would what they see make them feel more certain or more unsure?

That's the whole audit.

In the next section, we'll move from the idea of "visibility" to the thing that makes visibility trustworthy: communication that reinforces trust.

Not motivational, not loud—clear, stabilizing, and repeatable.

Communication That Reinforces Trust

A client's trust in you is not built only by what you do. It's built by what you say—and, just as importantly, how you say it.

Real estate is full of moments where the facts are not comforting. A low appraisal. A tough inspection report. A buyer who gets cold feet. A listing that doesn't get traction. A negotiation that drags. A rate change that shifts affordability.

In those moments, your tone becomes part of the service. Your clarity becomes part of the value. Your structure becomes part of the safety.

This is where a lot of agents unintentionally lose trust, even when they mean well. They either over-reassure ("Don't worry, it'll be fine") or over-inform (dumping details without interpretation). Neither creates stability. One creates false certainty; the other creates overwhelm.

Trust-building communication has a different signature. It doesn't perform confidence. It produces orientation. It helps the client understand what matters and what happens next.

The Stabilizing Structure

When you listen to the agents who clients describe as "calming," you'll notice they tend to communicate in a very consistent pattern. They don't speak in chaos. They speak in structure.

They do three things:

They name what's happening. They frame the tradeoff. They define the next step.

It's simple. It's repeatable. And it works because it turns uncertainty into something the client can hold.

Here's what this looks like in practice:

"This inspection item is common. Here's what it usually costs. Here's what we'll ask for, and here's what we'll do if they push back."

Or:

"The market is slower in this segment than it was last year. That doesn't mean we panic. It means we price with precision and negotiate with patience. Here's how we'll approach the next two weeks."

Or:

"We didn't win that offer. That doesn't mean we're failing. It means we learned the threshold for this neighborhood. Here's the adjustment."

The content is less important than the structure. The structure is what creates safety.

One of the simplest ways to reinforce trust is to set expectations early. Before you ever tour a property or sign a listing agreement, sit down with your clients and align on goals, timelines, and communication preferences. Ask them how often they want updates, what worries them, what success looks like, and how they prefer to receive information. Then commit to summarizing each conversation in writing so there are no surprises. Proactive expectation-setting reduces anxiety for your clients and gives you a clear framework to reference when challenges arise.

Proactive engagement goes hand in hand with expectation-setting. Don't wait for your clients to ask for updates; anticipate their questions. Share market intelligence, inspection status, or contract milestones before they have to ask. When you reach out first with strategic information, you replace anxiety with confidence and reinforce that you're on top of the details.

The Two Voices That Destroy Trust

When agents lose trust, it's often because they slip into one of two voices.

The Hype Voice

Everything is amazing, urgent, and absolute. This voice can create excitement, but it also triggers suspicion. When clients sense performance, they protect themselves.

Examples:

- "This is the BEST neighborhood in the entire city!"
- "You HAVE to act NOW or you'll lose it!"
- "This is a ONCE-IN-A-LIFETIME opportunity!"

The hype voice doesn't feel like guidance. It feels like sales. And in high-stakes decisions, sales language creates resistance.

The Anxious Voice

Everything is complex, uncertain, and spiraling. This voice may be honest, but without structure it transfers anxiety to the client. When clients absorb your anxiety, they lose confidence in the process—and in you.

Examples:

- "Well, it's really complicated, and honestly, I'm not sure what's going to happen..."
- "The market is so unpredictable right now, it's impossible to say..."
- "I've never seen a situation quite like this before..."

The anxious voice doesn't feel like leadership. It feels like shared panic.

The Trust Voice

The trust voice is different. It's calm without being passive. Clear without being dismissive. Direct without being dramatic.

It says: "We can handle this."

Examples:

- "Here's what happened. Here's what it means. Here's what we do next."
- "This is a challenge, but it's not uncommon. Here's how we address it."
- "You have two options here. Let me walk you through the tradeoffs of each."

The trust voice doesn't eliminate uncertainty. It organizes uncertainty into decisions.

And decisions are what clients need.

Making Your Communication Repeatable

The reason this matters beyond the transaction is that your communication becomes your reputation. People repeat what you said. They borrow your framing. They tell their spouse, "Our agent explained it like this..."

That repetition is how trust spreads.

It's also why you want your communication to be consistent across channels. The way you speak in a consultation should match the way you write a market note. Your website should feel like the same person who would guide a negotiation. When those pieces match, clients relax.

When they don't match, clients hesitate.

Field-Tested Principle

"Your clients don't need you to be a cheerleader. They need you to be a stabilizer. Give them structure, not hype. Give them clarity, not certainty. Give them the next step, not false promises."

Coaching Moment

Communication is your superpower—or your liability.

If you master the stabilizing structure, you become the agent people describe as "calm," "clear," and "trustworthy."

If you default to hype or anxiety, you become forgettable or stressful to work with.

The difference isn't personality. It's discipline.

Practice the three-part structure:

1. Name what's happening
2. Frame the tradeoff
3. Define the next step

Do that in every complex moment, and you'll build a reputation for competence that compounds over decades.

Your Website as Trust Confirmation

Most agents either overestimate what a website can do, or underestimate what it's quietly doing to them right now.

They overestimate it when they expect it to generate leads on autopilot with a few pages and a contact form. They underestimate it when they treat it like a digital business card—something that exists "because you're supposed to have one," but doesn't really matter.

In a trust economy, your website has one primary job:

It confirms the decision the client is already leaning toward.

It's not the spark. It's the seal.

When someone hears your name, or sees you in passing online, or gets referred by a friend, they don't always call immediately. They check. They're looking for reasons to feel safer. They want to verify that their instincts are correct. Your website is often where that verification happens.

And if your website doesn't confirm trust, it doesn't just fail to help—it can quietly create doubt.

The 5-second test: What visitors assess instantly

A modern client rarely sits down and reads your website like a brochure. They skim it like a risk assessment.

They are moving fast, often on a phone, with one question in the background: "Is this person safe?"

Here's what they assess in the first 5 seconds:

1. Is this person real? (Photo, name, contact info visible)
2. Are they active? (Recent content, current listings, proof of life)
3. Do they serve my area? (Geographic specificity, local references)
4. Do they seem competent? (Professional design, clear communication)
5. Can I contact them easily? (Phone number, email, contact form)

If any of these answers feels unclear or negative, trust leaks.

Not dramatically. Not consciously. But enough to create hesitation.

And in real estate, hesitation is expensive.

The client isn't "browsing." They're verifying.

Think about the last time you Googled a professional before hiring them.

You weren't reading every page. You were scanning for signals:

- Does this feel legitimate?
- Is this person current or stale?
- Do other people trust them?
- Does this match what I've heard?

Your website visitors are doing the same thing.

They're not looking for perfection. They're looking for coherence.

Coherence is when your presence tells a consistent story:

- You're active
- You're current
- You're specific to their area
- You communicate clearly
- Other people trust you

When coherence is present, clients feel a subtle relief. They may not say it out loud, but they think, "Okay. This makes sense."

When coherence is missing, they don't always reject you—they just hesitate.

And hesitation kills decisions.

The three things a trust-confirming website must do

A trust-confirming site doesn't need to be big. It needs to be coherent. It needs to do three things immediately.

1. Communicate who you help and where you help them (in plain language)

If a client can't tell within seconds whether you serve their area or their situation, they won't work harder. They'll bounce.

Bad example:
"Full-service real estate professional serving buyers and sellers."

Could be anyone. Could be anywhere.

Good example:
"I help first-time buyers and growing families find homes in East Dallas neighborhoods: Lakewood, M Streets, and Junius Heights. If you're trying to balance school districts, walkability, and budget, I can help."

Specific. Geographic. Clear client type.

2. Demonstrate how you think (without jargon)

A client should be able to read a few lines and feel, "This person explains things the way I want them explained."

That's competence trust in miniature.

Bad example:
"Leveraging market analytics and strategic positioning to maximize ROI."

Sounds impressive. Means nothing.

Good example:
"The market is slower than last year, but that doesn't mean panic. It means we price with precision and negotiate with patience. Here's how I guide clients through this shift."

Clear. Calm. Specific.

3. Provide proof (human proof, not boasts)

Not "Top Producer" badges.

Not stock photos of happy families.

Real proof:

- Client testimonials with names and specifics
- Recent closings with stories (not just addresses)
- Reviews from Google/Zillow
- Video testimonials (even better)
- Evidence of recent activity

When those three things are present, your website becomes a force multiplier. It quietly closes the loop for people who are already leaning toward you.

When they're missing, your website becomes a leak.

Where websites leak trust (12 specific issues + fixes)

Trust leaks almost never come from one catastrophic problem. They come from small inconsistencies that add up to a question mark.

Leak 1: Outdated content

What it signals: "Are they still in business?"

Fix: Update your blog/news section monthly minimum. If you can't maintain a blog, remove the section entirely.

Leak 2: Generic language

What it signals: "They could be anyone."

Fix: Replace template language with specific references to your market, your clients, your approach.

Leak 3: No recent activity

What it signals: "Are they active or coasting?"

Fix: Add "Recent Sales" or "Latest Content" section. Update monthly.

Leak 4: Missing or buried phone number

What it signals: "Do they actually want calls?"

Fix: Phone number in header, visible on every page. Make it click-to-call on mobile.

Leak 5: Slow load time (3+ seconds)

What it signals: "This person doesn't care about my experience."

Fix: Optimize images, use modern hosting, enable caching. Test on mobile.

Leak 6: Not mobile-optimized

What it signals: "This person is behind the times."

Fix: Use responsive design. Test on phone before launch.

Leak 7: Broken links or missing pages

What it signals: "Sloppy or abandoned."

Fix: Quarterly link check. Fix immediately.

Leak 8: Stock photos instead of real photos

What it signals: "Generic. Not personal."

Fix: Use your actual photo. Your actual neighborhood photos. Your actual office.

Leak 9: No proof of expertise

What it signals: "What makes you qualified?"

Fix: Add testimonials, reviews, case studies, recent closings, video testimonials.

Leak 10: Forcing contact before providing value

What it signals: "You just want my info."

Fix: Provide value first (neighborhood guides, market reports). Make contact optional, not required.

Leak 11: Too many CTAs (call-to-action buttons)

What it signals: "Desperate."

Fix: One primary CTA per page. Make it natural, not aggressive.

Leak 12: Unclear navigation

What it signals: "I can't find what I need."

Fix: Simple menu: Home, About, Neighborhoods, Resources, Contact. That's it.

Before/after examples: What trust confirmation looks like

Example 1: The generic agent

Before (trust leaks):

Homepage headline:
"Your Trusted Real Estate Partner"

About page:
"With years of experience and a commitment to excellence, I provide full-service real estate solutions for all your needs."

Recent activity:
Blog last updated 8 months ago

Contact:
"Send me a message" form. No phone number visible.

What the visitor thinks:
"This could be anyone. Are they even active? How do I actually reach them?"

After (trust confirmed):

Homepage headline:
"I help families find homes in West Plano with great schools, walkable neighborhoods, and strong resale value."

About page:
"I moved to Plano in 2015 with my own young family. I know what you're looking for: good schools without overpaying, neighborhoods where kids play outside, and homes that hold value. I've closed 120+ transactions in West Plano and know every street, every builder, every hidden issue to avoid."

Recent activity:
Blog post from this week: "Why West Plano home prices stayed flat in Q4 (and what it means for spring buyers)"

Contact:
Phone number in header: (214) 555-1234 - Click to call
"Call or text anytime. I answer my phone."

What the visitor thinks:
"This person knows my area. They understand what I need. I can actually reach them."

Example 2: The luxury agent

Before (trust leaks):

Homepage:
Slideshow of generic luxury interiors (stock photos)
Headline: "Luxury Real Estate Excellence"

About page:
"As a top producer in luxury real estate, I provide white-glove service and unparalleled market knowledge."

Listings:
Addresses only. No stories. No context.

What the visitor thinks:
"Impressive photos, but who are they? Do they actually work with people like me?"

After (trust confirmed):

Homepage:
Video walkthrough of recent $3M listing in Summerlin
Headline: "I represent buyers and sellers of luxury homes in Summerlin, Red Rock, and The Ridges."

About page:
"I've lived in Summerlin for 15 years and closed $140M in luxury transactions. My clients are typically established professionals, empty-nesters, or relocating executives who value discretion, expertise, and white-glove service. I don't just sell homes—I guide you through one of the most important decisions you'll make."

Listings:
Recent sale: "$2.8M custom estate in The Ridges - Represented seller who needed fast, discreet sale during relocation. Closed in 28 days, $50K over list."

What the visitor thinks:
"This person understands my market and my needs. They've done this before. I trust them."

Stan's principle: "It's nuts not to put your phone number down"

Stan has a pet peeve that tells you everything you need to know about how he thinks:

"Sometimes I see agents' websites, and there's no phone number. Just a form. Or it says, 'Send me a message.'"

"That's nuts."

"This is a house. This is probably the biggest financial decision someone is going to make. And you want them to fill out a form and wait for you to get back to them?"

"People want to talk. They want to hear your voice. They want to know you're real. They want to feel like if something goes wrong, they can reach you."

"So I put my phone number everywhere. On my website. In my emails. On my social profiles. I want people to know: if you need me, you can call me. Right now."

That's not a tactic. That's a trust philosophy.

Friction kills trust.

And when the stakes are high, people don't want barriers. They want access.

Field-Tested Principle

"Every extra step between a client and clarity is a chance for trust to leak."

Technical requirements (the basics you can't skip)

Your website doesn't need to be fancy. But it does need to meet basic technical standards.

Mobile optimization (non-negotiable)

60-70% of your traffic is mobile. If your site doesn't work on a phone, you're losing most visitors.

Test: Open your site on your phone. Can you:

- Read the text without zooming?
- Tap buttons without hitting the wrong one?
- Navigate easily?

If no to any of these, fix it immediately.

Load speed (under 3 seconds)

Google penalizes slow sites. Visitors bounce from slow sites.

Test: Use Google PageSpeed Insights (free tool)
Goal: 80+ score on mobile

Common fixes:

- Compress images
- Use modern hosting (not $5/month shared hosting)
- Enable caching
- Minimize plugins

SSL certificate (the padlock)

Sites without SSL (https://) trigger browser warnings.

Fix: Most hosting providers offer free SSL. Enable it.

Contact information visible

Phone, email, contact form—all should be easy to find.

Best practice: Phone number in header, visible on every page.

Metrics that matter (website edition)

Most agents track the wrong website metrics. They obsess over total traffic without understanding where it's coming from or what it's doing.

What to track:

1. Branded search trend (Google Search Console)
Are more people searching your name over time? That means awareness is growing.

2. Google Business Profile actions
Calls, direction requests, website clicks from your GBP listing.

3. Contact intent actions
Call button clicks, form submissions, email clicks.

4. Return visitor ratio
If 40%+ of your traffic is returning visitors, your content is valuable enough to bring people back.

5. Time on key pages
Are people reading your About page? Your neighborhood guides? Or bouncing immediately?

What to ignore:

1. Raw traffic without intent
10,000 visitors from a viral post ≠ 100 visitors searching for an agent in your area.

2. Bounce rate (mostly)
High bounce rate on a blog post isn't bad if the post answered their question. Context matters.

3. Total pageviews
Pageviews don't predict deals. Contact intent does.

The "trust confirmation stack" checklist

When a client lands on your website, they're moving through a mental checklist. You can't control the order, but you can make sure each element is present.

The sequence:

1. Real + Reachable
 ☐ Real photo of you
 ☐ Your actual name
 ☐ Phone number visible
 ☐ Local presence confirmed

2. Current
 ☐ Recent content (blog, news, updates)
 ☐ Recent listings or closings
 ☐ Current year mentioned

3. Specific Fit
 ☐ Geographic specificity (neighborhoods, areas)
 ☐ Client type clarity (who you serve)
 ☐ Examples that match visitor's situation

4. Clear Thinking
 ☐ At least one well-written page

☐ Helpful resource (not just sales pitch)

☐ Calm, professional tone

5. Proof

☐ Client testimonials (with names)

☐ Reviews visible or linked

☐ Outcomes shared (closings, success stories)

☐ Social proof (other people trust you)

When all five are present, the client moves from "maybe" to "probably" to "yes, let's talk."

Your website is not a billboard. It's a mirror.

Your website doesn't create trust. It confirms it.

When clients look into it, they should see the agent you actually are—calm, clear, competent, current.

If they see something generic, stale, or mismatched, they hesitate.

And hesitation costs you deals.

Coaching Moment

If a stranger looked you up tonight—right now—would what they see make them feel more certain or more unsure?

That's the whole audit.

Your website isn't about impressing people. It's about removing doubt.

Make sure that when someone checks, what they see increases confidence instead of creating questions.

END OF PART V

Part V Summary & Transition

We've made trust legible:

1. Visibility Without Credibility Is a Liability — Coherence beats volume
2. Communication That Reinforces Trust — Structure creates safety (name it, frame it, define next step)
3. Your Website as Trust Confirmation — It's not the spark; it's the seal

Now we shift from making trust visible to protecting it over time.

In Part VI, we'll build the systems that prevent trust drop-off and help you adapt without fragmenting—so you can remain relevant through every market cycle.

PART

VI

DURABILITY: PROTECTING THE EDGE

The Trust Drop-Off Problem

You've built trust. You've shown up consistently. You've earned referrals. You've closed deals. Your sphere knows you're an agent. Your past clients are happy.

And then life happens.

You get busy. You have three closings at once. Your kid gets sick. You take a vacation. A family emergency pulls you away. The busy season hits and you're heads-down in transactions for eight weeks straight.

When you finally come up for air, you realize something uncomfortable:

Your email list hasn't heard from you in two months. Your social media went dark. Your website blog hasn't been updated. Your sphere forgot you were still active.

And now your pipeline feels thin.

This is the Trust Drop-Off Problem.

It's not that people stopped liking you. It's that your relevance cooled while you were working.

And in a trust economy, relevance decay is expensive.

Why Busy Agents Disappear (And Why It Costs Them)

The irony is brutal: the agents who disappear most often are the ones who are busiest serving clients.

They're doing excellent work. They're delivering value. They're living their standards.

But while they're heads-down executing, their visibility goes silent. And silence is interpreted as absence.

The market doesn't experience your busyness as "they're working hard."

It experiences it as "I haven't heard from them in a while."

And when that silence extends for weeks or months, something subtle but damaging happens:

You slip from "top of mind" to "I should probably check in with them sometime."

And "sometime" rarely comes before "right now I need an agent."

This is why consistency compounds—and why interruption is so expensive.

When you disappear for a quarter, you don't just lose three months of visibility. You lose the compounding momentum of the previous year.

Stan's Story: "I Learned This the Hard Way"

Stan remembers a season early in his career when business was good—too good.

He had multiple closings happening simultaneously. Inspections, appraisals, negotiations, title issues—all demanding his attention at once.

"I was working twelve-hour days. I was delivering for my clients. I was doing everything right inside the transactions."

"But I stopped posting. I stopped sending my market updates. I stopped touching my sphere. I told myself, 'I'll get back to it when things slow down.'"

"And when things finally slowed down, I realized my pipeline was empty. I had no new leads. My referrals had dried up. People in my sphere were working with other agents—not because they didn't like me, but because they forgot I was active."

"I learned something that day: Your visibility can't be seasonal. It has to survive your busy seasons, or you'll always be rebuilding."

That lesson changed how Stan operates.

He built systems—not just tactics—that protect his presence even when life gets full.

He automated reminders. He batched content. He used AI to maintain his email cadence. He built a minimum viable presence that didn't require him to be "on" every day, but ensured he never went silent for months.

"Now, even when I'm slammed, my sphere hears from me. My social media stays active. My email goes out. It's not always my best work, but it's consistent. And consistency is what keeps the phone ringing."

Field-Tested Principle

"Your visibility must survive your busy seasons, or you'll spend your slow seasons rebuilding instead of closing."

The Minimum Viable Presence (MVP)

The solution to the Trust Drop-Off Problem is not "never get busy."

The solution is building a Minimum Viable Presence—a baseline level of visibility that continues even when your attention is needed elsewhere.

MVP is not about being everywhere or posting daily. It's about having a floor below which you never drop, regardless of how full your schedule gets.

Here's what MVP looks like in practice:

Email: One message per month to your sphere (minimum)

- Can be batched in advance
- Can be AI-assisted (with your review)
- Template-based but personalized

Social Media: 2-3 posts per week

- Batched during slow periods
- Repurposed across platforms
- Mix of value, proof of life, and humanity

Google Business Profile: Updated monthly

- New photo
- Post or update
- Active review responses

Website: Quarterly refresh minimum

- New blog post or market update
- Verify all info is current
- Check for broken links

This is not ambitious. This is protective.

It's the difference between "I'm building momentum" and "I'm preventing decay."

The Batching Strategy

The secret to maintaining MVP during busy seasons is batching—creating content in advance during your slower periods.

Most agents do the opposite. They create content when they feel inspired or when they have time, which means their output is erratic and unpredictable.

Market leaders batch.

Here's how:

Monthly Content Batching Day (3-4 hours):

- Record 4-5 videos (one per week)
- Write 2-3 email drafts
- Capture 10-15 social media photos
- Draft 8-12 social posts

Quarterly Deep Batch (Half day):

- Plan content themes for next 90 days
- Batch record 12-15 videos
- Write 6 email templates
- Create social media content calendar

When you batch, you're not creating content in real-time. You're scheduling consistency in advance.

Then, when you get busy, your presence continues without requiring daily effort.

Why "Just Post More" Doesn't Work

Some agents hear "stay consistent" and interpret it as "post daily."

That's not the answer. Daily posting without purpose doesn't build trust—it creates noise.

The market doesn't reward frequency. It rewards coherence + consistency over time.

Three valuable posts per week, maintained for 52 weeks, beats daily posting for 6 weeks followed by silence for 6 months.

This is why MVP is powerful: it's sustainable. It doesn't require you to be "on" all the time. It requires you to have systems that protect your floor.

The Four Systems That Prevent Trust Drop-Off

If you want to protect against relevance decay, you need four systems in place:

System 1: Content Production System
Batching schedule, templates, AI assistance, repurposing framework

System 2: CRM Maintenance System
Weekly database hygiene, automated follow-up reminders, segmentation protocols

System 3: Communication Calendar
Monthly email schedule, social posting rhythm, quarterly touchpoints for VIPs

System 4: Accountability System
Weekly review of metrics (emails sent, posts published, contacts touched), monthly audit of presence (website, GBP, social profiles)

These aren't complicated. But they must be explicit.

Because when life gets full, the only things that survive are the things you've systematized.

Stan's Approach: "I Schedule It Like a Closing"

Stan treats his marketing system the way he treats a closing: non-negotiable.

"I have a recurring block on my calendar every Friday: 'Marketing Hour.' It's sacred. I don't book appointments during it. I don't let other things creep in."

"During that hour, I review my database, I send follow-ups, I check my content calendar, I batch a few posts for next week. That's it."

"One hour per week keeps my presence alive. And when I get busy, that one hour is what prevents me from disappearing."

That discipline—treating visibility as infrastructure, not inspiration—is what separates agents who compound trust from agents who cycle through chaos.

Field-Tested Principle

"Consistency isn't about intensity. It's about having a floor you never drop below, no matter how busy you get."

The Reactivation Protocol

Even with systems, there will be times when you slip. Life happens. Emergencies happen. You go silent for longer than you planned.

When that happens, don't spiral. Don't feel guilty. Don't ghost your sphere out of embarrassment.

Reactivate.

Here's the simple reactivation protocol:

Step 1: Acknowledge the gap (briefly)
"Hey everyone—it's been a minute. I've been deep in closings and life, but I'm back."

Step 2: Provide immediate value
Share a market update, a neighborhood insight, or a helpful resource. Give them a reason to re-engage.

Step 3: Recommit to consistency
"You'll be hearing from me regularly again—I've missed staying connected."

Step 4: Rebuild momentum
Get back to your MVP rhythm. Don't try to compensate with a burst of activity. Just resume your baseline.

Most agents overcomplicate reactivation. They think they need to apologize profusely or create some grand comeback campaign.

You don't.

You just need to show up again, provide value, and stay consistent going forward.

Coaching Moment

The Trust Drop-Off Problem is not a character flaw.

It's a systems gap.

If your visibility can't survive your busy seasons, you don't need more discipline—you need better systems.

Build the MVP. Batch your content. Protect your floor.

Because the agents who win long-term aren't the ones who never get busy.

They're the ones whose presence compounds even when they're executing.

Adapting Without Fragmenting

Markets change. Platforms evolve. Client preferences shift. Economic cycles turn. Technology advances.

The agents who survive these changes are not the ones who resist adaptation.

They're the ones who adapt without fragmenting.

Fragmentation is what happens when you chase every new tactic, platform, or trend without a stable core. You start strong on Instagram, then switch to TikTok, then try YouTube, then abandon all three for the next thing. You rewrite your positioning every six months. You change your voice to match whatever style seems to be working for someone else.

And in the process, you lose the one thing you were trying to build: a coherent, recognizable presence that people trust.

This section is about how to stay relevant without losing your identity.

The Stable Core / Flexible Surface Model

The mental model that protects you is simple:

Stable Core: The things that don't change

- Your values
- Your positioning (who you serve, where, why)
- Your communication philosophy
- Your standards

Flexible Surface: The things that can adapt

- Platforms (Instagram, TikTok, YouTube, whatever's next)
- Content formats (Reels, Stories, long-form, podcasts)
- Tools (CRM, AI, automation tech)
- Tactics (SEO, ads, email sequences)

When your core is stable, you can experiment with the surface without feeling like you're starting over every time something changes.

You can try TikTok without abandoning your voice. You can test AI tools without losing your judgment. You can shift platforms without confusing your market.

Because your market isn't following you for the platform. They're following you for the clarity, the consistency, and the competence you represent.

The platform is just the delivery mechanism.

Why Agents Fragment

Fragmentation usually happens for one of three reasons:

Reason 1: FOMO (Fear of Missing Out)
A new platform explodes. An agent sees someone else crushing it. They panic and shift all their energy to the new thing, abandoning what was already working.

Reason 2: Impatience
They try something for 60 days, don't see immediate results, and pivot to the next tactic. They never stay with anything long enough for it to compound.

Reason 3: No Clear Positioning
Without a stable core, every new trend feels like it might be "the answer." They chase because they don't have a center to return to.

The antidote to all three is the same: clarity on what doesn't change.

When you know your core, adaptation becomes easy. You're not reinventing yourself. You're just finding new ways to express the same consistent thing.

Stan's Story: "I Don't Chase Platforms. I Use What Works for My People."

Stan has watched platforms rise and fall over his 22-year career.

"When I started, it was all about websites and SEO. Then Facebook exploded. Then Instagram. Then video became the thing. Now it's AI, TikTok, Reels—there's always something new."

"Early on, I would stress about it. I'd think, 'If I'm not on the new platform, I'm going to fall behind.' So I'd jump on every new thing, try to figure it out, feel overwhelmed, and then abandon it."

"Now I do it differently. I ask one question: Where are my people?"

"My best clients—the ones with equity, the ones who refer, the ones I actually want to work with—they're on Facebook and email. Some are on Instagram. Almost none are on TikTok."

"So I focus there. I don't chase trends. I serve my market."

"And when a new tool or platform actually becomes relevant to my clients—when I see evidence they're using it—then I'll learn it. But I don't panic just because something's trending."

That's the difference between adapting strategically and fragmenting reactively.

Field-Tested Principle

"Don't chase platforms. Serve your people. When your people move, you move. Until then, double down on what's working."

The 90-Day Test

If you want to experiment with a new platform, tool, or tactic without fragmenting, use the 90-Day Test.

Here's how it works:

Before you start:

- Define what success looks like (not "go viral"—real metrics: engagement, DMs, calls, referrals)
- Commit to 90 days of consistent effort

- Don't abandon what's already working

During the 90 days:

- Show up consistently (doesn't have to be daily, but must be regular)
- Track leading indicators (replies, engagement, branded searches)
- Give it enough time to see if it compounds

At 90 days:

- Evaluate: Is this creating real relationships or just noise?
- Decide: Keep, modify, or cut
- Don't immediately jump to the next thing

This prevents you from platform-hopping every 30 days while still giving you space to experiment intelligently.

How to Know When to Pivot vs. When to Persist

Sometimes persistence is smart. Sometimes it's stubborn.

Here's how to tell the difference:

Persist when:

- You're seeing slow but consistent growth in leading indicators (engagement, replies, DMs, branded searches)
- Your best clients say, "I've been seeing your stuff"
- The work feels aligned with your positioning

- You haven't given it enough time to compound (most things need 6-12 months)

Pivot when:

- You've given it 6+ months and see zero engagement or real relationship growth
- The platform doesn't match where your ideal clients actually are
- It's draining you without producing any business or referral momentum
- You're becoming someone you're not just to fit the platform

The key: Don't pivot out of impatience. Pivot based on evidence.

The Technology Adoption Framework

New tools will always emerge. AI will evolve. CRMs will add features. Platforms will launch.

Here's how to evaluate whether to adopt something new:

Ask four questions:

1. Does this solve a real problem I have? (Not a hypothetical problem—a real one)
2. Does this align with my stable core? (Does it help me serve my people better, or does it pull me off-brand?)
3. Can I sustain this alongside what's already working? (Or will it fragment my effort?)
4. What's the cost of waiting 6 months to see if it sticks? (FOMO vs. real opportunity cost)

If the answers are yes, yes, yes, and "low," consider testing it.

If not, bookmark it and revisit in six months.

Field-Tested Principle

"The graveyard of real estate careers is filled with agents who chased every trend and mastered none. Depth beats breadth. Consistency beats novelty. Stable core beats flexible identity."

Coaching Moment

Markets will change. Platforms will rise and fall. Tools will evolve.

Your job is not to predict the future.

Your job is to build a foundation so stable that you can adapt to whatever comes without losing yourself in the process.

When your positioning is clear, your communication is consistent, and your systems are sound, you can experiment freely.

Because you're not betting your identity on the experiment.

You're just testing a new delivery mechanism for the same core value you've always provided.

That's how you stay relevant for decades instead of chasing relevance every quarter.

The 90-Day Launch Plan

You've read the book. You understand the philosophy. You see the frameworks. You know what matters.

Now the question is: Where do you start?

Most agents make one of two mistakes:

1. They try to implement everything at once, get overwhelmed, and quit in 30 days.
2. They finish the book feeling inspired but never actually start because they don't know what to do first.

This is where we solve both problems.

This is your implementation roadmap—a focused 90-day plan that moves you from concept to execution without fragmenting your effort.

The Philosophy Behind the Plan

This plan is built on three principles:

Principle 1: Sequential, Not Simultaneous
You will not try to master email, social, video, CRM, AI, and websites all at once. You will build one layer at a time, letting each layer stabilize before adding the next.

Principle 2: Foundation First, Tactics Second
You will not start with "posting more." You will start with clarity

—positioning, voice, standards—so that when you do create content, it's coherent instead of generic.

Principle 3: Minimum Viable Presence, Then Scale
You will not aim for perfection. You will aim for consistency at a sustainable baseline. Once that baseline is stable, you can scale.

The 90-Day Structure

Weeks 1-2: Foundation (Clarity & Systems)
Weeks 3-6: Channel One (Master Your Primary Channel)
Weeks 7-10: Channel Two (Add Your Secondary Channel)
Weeks 11-12: Integration & Optimization

Let's break it down.

WEEKS 1-2: FOUNDATION

Goal: Get clear on positioning, voice, and baseline systems before creating any content.

Week 1 Tasks:

Day 1-2: Run Your Trust Audit

- Google yourself. What do you see?
- Check your website, social profiles, Google Business Profile
- Identify gaps: What signals are missing? What feels stale or generic?

Day 3-4: Clarify Your Positioning

- Who do you serve? (Client type)
- Where do you serve them? (Geographic specificity)

- What are you known for? (The outcome or edge you deliver)
- Write one sentence: "I help [WHO] in [WHERE] [DO WHAT]."

Day 5-7: Define Your Voice & Standards

- Write down 5 phrases you'd never say (hype, clichés, pressure language)
- Write down 3 communication principles you always follow (clarity, calm, specificity, etc.)
- Find 2-3 examples of past communication you're proud of (emails, posts, explanations)

Week 2 Tasks:

Day 1-3: CRM Setup

- Choose your CRM (if you don't have one: start with HubSpot free or Follow Up Boss)
- Import your database
- Segment into tiers: Past Clients, Sphere, Active Leads, Warm Prospects, Cold
- Tag contacts by source, interest, geography

Day 4-5: Build Your Context Packet (AI Training)

- Compile: Your positioning, your voice rules, your communication examples
- Store in one document that you can feed to AI when you need help drafting
- Test it: Ask AI to draft a market update using your Context Packet. Does it sound like you? Refine.

Day 6-7: Set Up Your Communication Calendar

- Decide: Which platforms will you use? (Pick 1-2 max for now)
- Decide: What's your email frequency? (Start with monthly minimum)
- Block recurring time: 1 hour per week for content batching/CRM maintenance

Outcome by end of Week 2:
You have clarity on who you are, who you serve, and how you communicate. Your systems are set up. You're ready to create content that's coherent instead of generic.

WEEKS 3-6: CHANNEL ONE (MASTER YOUR PRIMARY CHANNEL)

Goal: Build consistency on ONE channel before adding more.

Choose Your Channel:

If you're comfortable on camera: Start with video (YouTube + repurpose to Instagram/Facebook)

If you're better in writing: Start with email (monthly value emails to your sphere)

If you love social but hate video: Start with Instagram or Facebook (70-20-10 content framework)

Week 3-4 Tasks: Content Creation

If Video:

- Script 4 neighborhood tour or market update videos
- Batch record all 4 in one session (use phone, natural light, tripod)
- Edit simply (or hire someone on Fiverr for $20/video)

- Upload to YouTube with optimized titles/descriptions
- Repurpose to Instagram Reels and Facebook

If Email:

- Write 4 monthly value emails (use templates + AI to draft, then refine)
- Schedule them in your email platform (one per week for next month)
- Focus on: market insights, neighborhood spotlights, helpful resources, personal updates

If Social Media:

- Plan 12 posts using 70-20-10 framework:
  - 8-9 value/community posts
  - 2-3 education/authority posts
  - 1-2 business posts
- Batch create (photos, captions) in one 2-hour session
- Schedule using Meta Business Suite or Later

Week 5-6 Tasks: Engagement & Refinement

- Publish your content consistently (1x/week minimum)
- Spend 15 min/day engaging: reply to comments, comment on others' posts, send DMs
- Track what's working: Which posts get replies? Which emails get responses?
- Refine: Adjust your voice, topics, format based on what resonates

Outcome by end of Week 6:
You have 4-6 weeks of consistent content on one channel. You're starting to build momentum. Your market is seeing you regularly.

WEEKS 7-10: CHANNEL TWO (ADD YOUR SECONDARY CHANNEL)

Goal: Layer in a second channel without abandoning the first.

Your Options:

- If you started with email → Add social media (repurpose your email content into posts)
- If you started with video → Add email (send your videos to your sphere monthly)
- If you started with social → Add email (turn your best posts into email content)

Week 7-8 Tasks:

- Plan your second channel's content (repurpose from Channel One wherever possible)
- Batch create 4 weeks of content
- Set up scheduling/automation

Week 9-10 Tasks:

- Publish consistently on BOTH channels
- Maintain CRM hygiene (weekly database touches, follow-ups)
- Continue engagement (replies, comments, DMs)

Outcome by end of Week 10:
You're now consistently present on two channels. Your sphere is hearing from you regularly. Your visibility is protected even when you get busy.

WEEKS 11-12: INTEGRATION & OPTIMIZATION

Goal: Tie everything together and identify what to improve.

Week 11 Tasks: Integration

Day 1-2: Website Audit

- Does your website reflect your current positioning?
- Is your content current?
- Are your contact details easy to find?
- Make updates as needed

Day 3-4: Google Business Profile Optimization

- Add fresh photos
- Post an update
- Respond to reviews
- Verify all info is accurate

Day 5-7: CRM Deep Clean

- Review all contacts: Are tags accurate?
- Set up automated follow-up sequences for warm prospects
- Plan next quarter's touchpoints for VIPs

Week 12 Tasks: Optimization & Planning

Day 1-3: Review Metrics

- Email: Open rates, reply rates
- Social: Engagement rate, DMs, branded searches
- CRM: Contacts touched, deals in pipeline

- Identify: What's working? What's not?

Day 4-5: Refine Your Systems

- Adjust your content calendar based on what resonated
- Update your Context Packet if your voice has evolved
- Tighten your batching process

Day 6-7: Plan Next 90 Days

- What will you keep doing?
- What will you test?
- What will you drop?
- Set goals for next quarter

Outcome by end of Week 12:
You have a stable, repeatable system. You're showing up consistently on two channels. Your database is organized. Your presence is protected. You're ready to scale or refine.

The Weekly Routine (Weeks 3-12)

Here's what your weekly rhythm should look like once you're in execution mode:

Monday (15 min): Review week's content calendar, confirm what's going out

Tuesday (15 min): Engage on social media (10-15 meaningful comments/DMs)

Wednesday (15 min): CRM maintenance (log interactions, set follow-up tasks)

Thursday (15 min): Engage + respond to any DMs/comments from your content

Friday (60 min): Batch next week's content OR deep CRM work (alternate weeks)

Total time: ~2 hours per week

That's it.

Two hours per week protects your presence, keeps your database warm, and ensures you don't disappear even when you're slammed.

What Not to Do

Don't try to be on every platform. Pick two. Master them.

Don't aim for perfection. Aim for consistency. Your 10th post will be better than your 1st.

Don't skip the foundation. If you jump straight to tactics without clarity, you'll create noise instead of trust.

Don't compare your Day 10 to someone else's Day 1,000. They've been at this for years. You're just starting. Stay in your lane.

Don't abandon what's working to chase the new thing. Test carefully. Don't fragment.

Resources & Support

Everything mentioned in this book—templates, worksheets, AI training guides, content calendars, CRM setup checklists—is available at:

therealtorlab.com/edge

You don't need them to implement this plan, but they'll make execution faster and easier.

Think of the book as the why and the what. The Lab is the how.

Final Thought on Implementation

The difference between agents who finish this book and feel inspired versus agents who finish this book and actually change their business is simple:

Action beats intention.

You don't need to do everything perfectly. You need to start.

Pick your first channel. Batch your first four pieces of content. Send them. Engage. Adjust. Repeat.

Do that for 90 days, and you'll have built something real.

Do that for a year, and you'll have built something durable.

Do that for a decade, and you'll have built the Realtor Edge.

Now go build it.

Long-Term Trust: The Compounding Advantage

Most agents think in transactions.

Market leaders think in decades.

This final section is about what happens when you stop chasing deals and start building equity in relationships—the kind of equity that doesn't reset when the market shifts, when platforms change, or when competitors flood in.

This is about the compounding advantage: the unfair edge that comes from doing simple things consistently over time until they become impossible to replicate.

The Math of Compounding Trust

Imagine two agents, both starting with 100 people in their sphere.

Agent A: Posts sporadically. Sends emails when they need business. Ghosts their sphere when they're busy. Shows up with energy during slow seasons, then disappears during busy ones.

Over five years, their sphere stays roughly the same size. They're always rebuilding. They close 8-12 deals per year, mostly from people who randomly decided to move and happened to remember them.

Agent B: Maintains MVP. Sends monthly emails. Posts 3x per week. Engages consistently. Never goes silent for more than two weeks, even during their busiest seasons.

Over five years, their sphere grows—not because they're chasing strangers, but because happy clients refer them. Their business becomes 70% referral-based. They close 20-30 deals per year, with less stress, higher margins, and better client fit.

The difference isn't talent. It's compounding.

Agent A resets every year. Agent B compounds.

And compounding is exponential, not linear.

In Year 1, the difference might be 2-3 extra deals. By Year 5, it's a completely different business.

Stan's Story: "$400K from One Relationship"

Stan traces $400,000 in commission back to one relationship.

Not one transaction. One relationship.

"I did a deal with a guy at the country club. It wasn't huge—a nice home, but nothing crazy. But I treated him like family. I stayed in touch. I sent him market updates. I checked in a few times a year. Just stayed present."

"A few years later, he called me. He wanted to flip a house. Then another. Then he introduced me to his brother. Then his business partner. Then a friend who was moving to Vegas."

"Over the course of a decade, that one relationship turned into thirty transactions. Land deals, investment properties, luxury homes, flips—everything."

"None of that happens if I close the first deal and disappear. None of it."

That's the compounding advantage.

One relationship, nurtured consistently, turned into a portfolio of business that funded years of Stan's career.

Field-Tested Principle

"Transactions are linear. Relationships are exponential. Choose accordingly."

Why Compounding Feels Slow (And Why It's Worth It)

Here's the uncomfortable truth about compounding: it feels slow at first.

In the early months, you're putting in effort and not seeing proportional returns. You're sending emails that get few replies. You're posting content that gets modest engagement. You're staying consistent when it feels like no one's paying attention.

And that's exactly when most agents quit.

They think, "This isn't working," and they pivot to the next tactic.

But they're wrong.

It is working. It's just not visible yet.

Because trust doesn't compound in a straight line. It compounds like this:

- Months 1-6: Minimal visible results. You're planting seeds.
- Months 7-12: Occasional wins. People start noticing.
- Months 13-24: Momentum builds. Referrals increase. Your name comes up more often.

- Years 3-5: Exponential returns. You're the obvious choice in your lanes. Referrals become your primary source of business.

The agents who win long-term are the ones who stay consistent through the slow part.

They understand that the effort they're putting in today is building equity they'll benefit from for years.

The Invisible Equity You're Building

When you show up consistently—when you stay present, stay useful, stay clear—you're building assets that don't show up on a balance sheet but are worth more than most visible assets.

You're building:

Top-of-Mind Equity: When someone thinks "real estate," they think of you first.

Referral Equity: Your past clients become your sales team. They refer you without you asking.

Reputation Equity: Your name becomes associated with clarity, competence, and calm.

Content Equity: Your body of work—videos, emails, posts—continues to work for you long after you create it.

Relationship Equity: The depth of trust you've built with your sphere can't be replicated by a new agent, no matter how good their ads are.

These assets appreciate over time. And they're defensible.

A competitor can copy your website. They can mimic your social media. They can run better ads.

But they can't replicate a decade of consistent, valuable presence in your market.

That's your moat.

What This Looks Like in Year 10

If you implement the principles in this book—clarity, consistency, systems, AI as leverage, legibility—and you maintain them for a decade, here's what your business looks like:

- 70-80% of your business comes from referrals
- Your cost per deal is a fraction of what other agents spend on lead gen
- You rarely "hunt" for business—it finds you
- Your clients already trust you before the first conversation
- You can take a two-week vacation without your business collapsing
- Your pipeline stays full even when you're not actively marketing
- You're known in your market as "the agent who explains things well"
- Your sphere refers you confidently because they've seen you show up for years

That's not hype. That's what compounding looks like in practice.

And the only way to get there is to start today and stay consistent.

The "Decade Agent" vs. "Deal Agent" Mindset

There are two types of agents:

Deal Agents: Think in transactions. Chase leads. Optimize for short-term wins. Celebrate closings but don't build systems to keep the relationship warm.

Decade Agents: Think in relationships. Nurture trust. Optimize for long-term compounding. Celebrate closings and immediately move the client into long-term nurture.

Deal Agents work hard every year because they reset every year.

Decade Agents work hard early, then benefit from compounding momentum.

The shift from Deal Agent to Decade Agent is simple:

Ask yourself after every closing: "How do I make sure this person—and everyone they know—thinks of me five years from now?"

Then build systems to make that inevitable.

The Real Realtor Edge

We've covered a lot in this book:

- Trust as the new currency
- Expertise that compounds through teaching
- Systems that protect consistency
- AI as an engine, not a replacement
- Making trust legible
- Adapting without fragmenting

But if you take only one thing from this book, take this:

The Realtor Edge is not a tactic. It's a commitment to being present, clear, and valuable over time—until trust becomes automatic.

You don't win by being louder than everyone else.

You win by being clearer, more consistent, and more useful than anyone else in your lane.

You win by staying when others disappear.

You win by teaching when others are selling.

You win by building systems that protect your presence when life gets full.

You win by compounding relationships while others chase transactions.

That's the edge.

And it's available to every agent willing to do the work.

Stan's Final Thought

"I've been doing this for over twenty years. I've seen markets boom and crash. I've seen platforms rise and fall. I've seen tactics that worked for a season and then stopped working.

"But the one thing that has never changed—the one thing that still works today the same way it worked in 2004—is this:

"If you treat people like family, if you stay present, if you make things make sense, and if you keep showing up even when no one's asking—you'll build a business that lasts.

"That's not sexy. It's not a hack. It's not a shortcut.

"But it works.

"And twenty years from now, when most of the agents who started with you have burned out or quit, you'll still be here.

"Not because you were the smartest. Not because you were the loudest.

"Because you were consistent.

"And consistency always wins."

Chase's Final Thought

"Real estate is one of the last industries where relationships still matter more than algorithms.

"Where trust still beats hype.

"Where clarity still beats volume.

"Where being yourself—clearly, consistently, at scale—still creates durable advantage.

"The tools will change. The platforms will evolve. The tactics will cycle.

"But the principles in this book won't.

"Because they're built on how humans actually choose, how trust actually forms, and how competence actually compounds.

"So use the tools. Build the systems. Leverage AI where it helps.

"But never forget: you're not building a content machine.

"You're building a trusted name in your market.

"And that takes time, discipline, and a willingness to show up when it's easier to disappear.

"Do that, and you won't just survive the next market cycle.

"You'll thrive through all of them."

CONCLUSION

The Work Ahead

You've reached the end of the book, but the beginning of the work.

The concepts are yours. The frameworks are clear. The plan is laid out.

Now it's on you to execute.

Start with the 90-Day Plan. Build your foundation. Master one channel. Add the second. Refine your systems. Protect your consistency.

And a year from now, when you're closing referral deals from relationships you nurtured, when your sphere is telling their friends about you, when your pipeline stays full even when you're not actively hunting—you'll understand what we mean by the Realtor Edge.

It's not loud. It's not flashy.

It's just durable.

And in a world that rewards noise, durability is the real unfair advantage.

Now go build it.

About the Authors

Stanley King

Stanley King has been a licensed real estate professional in Las Vegas, Nevada for over two decades. Since entering the industry in 2004, he has closed more than 1,500 transactions across nearly every segment of the market—luxury estates, first-time buyer homes, land deals, investment properties, foreclosures, and probate sales.

Stan's career has spanned multiple market cycles, including the boom years, the Great Recession, and the post-pandemic surge. Through each cycle, his business has remained stable—not because he chased trends, but because he built relationships that lasted.

His approach is grounded in a simple philosophy: treat clients like family, stay present in their lives, and make complex decisions feel manageable. That consistency has created a referral-based business where 70% of his deals come from past clients and their networks.

Outside of real estate, Stan is an avid outdoorsman and adventure enthusiast. He has raced the Baja 1000 twice—one of the most grueling off-road endurance races in the world—and maintains a YouTube channel documenting his dirt bike adventures, which has grown to over 20,000 subscribers. His willingness to share his passion authentically, without performance or perfection, mirrors the philosophy he brings to real estate: show up, be real, stay consistent.

Stan lives in Las Vegas with his family and continues to serve clients with the same standards he brought to his first transaction twenty years ago.

Chase Powell

Chase Powell is a serial entrepreneur, systems architect, and advocate for using technology to amplify human judgment rather than replace it.

He began his career in real estate in 2003, earning his license and working as an agent before recognizing that the industry was undergoing a fundamental shift in how clients discovered, evaluated, and chose professionals. Rather than resist that shift, he studied it—learning the mechanics of digital marketing, content strategy, search behavior, and CRM systems long before they became standard in real estate.

Over the past two decades, Chase has founded multiple companies in the digital marketing and AI space, including:

- CodeTap (codetap.ai) — AI-powered automation and workflow solutions
- Lubbock Web Guy (lubbockwebguy.com) — Web design and digital strategy
- The Agency Guild (agency.codetap.ai) — Resources and systems for marketing agencies
- PatientFlow (patientflow.us) — Med Spa & Clinic Ad Management & Digital Marketing

His work has helped real estate teams, small businesses, and entrepreneurs build digital infrastructure that produces trust, not just traffic. He specializes in translating complex systems into frameworks that non-technical professionals can implement and maintain.

Chase is also the author of Always Be Learning: The Pursuit of Knowledge in a World That Has Forgotten Its Power, a book exploring the discipline of continuous learning and intellectual growth. He is a speaker, teacher, mentor, and coach, passionate about helping people level up their lives through business, theology, health, and knowledge.

He is a traditional Roman Catholic, a member of the Iron Oak Society (a men's group focused on becoming better fathers, husbands, and businessmen), and an advocate for mentorship in all its forms. His hobbies include theology, cigars, scuba diving (he is a master scuba diver), traveling, live music, lifting weights, breathing techniques, gratitude journaling, and the outdoors.

Chase has been married for 23 years and has five children, ages 13 to 22. He lives in Lubbock, Texas, where he continues to build businesses, mentor emerging entrepreneurs, and pursue knowledge and truth in all areas of life.

About The Realtor Lab

The Realtor Lab exists to help real estate professionals build trust at scale without losing their identity.

We don't sell leads. We don't promise overnight results. We don't treat agents like they need to become influencers or marketers.

Instead, we provide the systems, tools, and frameworks that let you show up consistently, communicate clearly, and make your competence legible—so when timing activates, you're the obvious choice.

What We Provide

Templates & Frameworks

Email sequences, social media content calendars, video scripts, CRM workflows, and AI training guides—all designed to protect your voice while making execution faster.

AI Training Systems

Context Packets, prompt libraries, and voice calibration tools that let you use AI as leverage without sounding generic.

Implementation Support

Step-by-step guides, checklists, and walkthroughs that turn the concepts in this book into repeatable systems you can run every week.

Community & Accountability

A network of agents committed to building durable businesses through trust, not tactics. Monthly calls, shared wins, honest feedback.

Our Philosophy

We believe:

- Trust is built through consistency, not intensity
- Expertise compounds when you teach, not just transact
- Systems protect presence when life gets full
- AI is an engine, not an identity
- Specificity beats generality
- Clarity beats volume
- Relationships beat algorithms

If that aligns with how you want to build your business, we built The Realtor Lab for you.

Get Started

Visit therealtorlab.com/edge to access:

- The 90-Day Launch Plan (downloadable)
- Content templates and batching calendars
- AI Context Packet builder
- CRM setup checklists

- Trust Audit worksheet
- Video scripting templates
- Email sequence library

Everything mentioned in this book is organized, accessible, and ready to use.

The book gives you the why and the what.
The Lab gives you the how.

APPENDIX A
The Trust Audit Worksheet

Purpose: Evaluate how clients experience your presence when they check you online.

Instructions: Answer honestly. This is not about judgment—it's about diagnosis.

SECTION 1: FINDABILITY

1. When someone Googles your name, what appears in the first 5 results?

- My website
- My Google Business Profile
- My social media profiles
- Reviews or testimonials
- Brokerage profile
- Nothing / Outdated results / Wrong person

2. Is your Google Business Profile:

- Fully completed (photo, description, hours, website link)
- Partially completed
- Not claimed or verified

- I don't know

3. Do your social media profiles clearly state:

- Your name
- "Realtor" or "Real Estate Agent"
- Your city/market
- How to contact you

Score Section 1:
If you checked all boxes in questions 1-3: Strong findability
If you're missing 2-3 elements: Moderate findability
If you're missing 4+ elements: Weak findability

SECTION 2: CURRENCY (Are You Current?)

4. When was your last social media post?

- Within the past week
- 1-2 weeks ago
- 2-4 weeks ago
- 1-3 months ago
- 3+ months ago

5. When was your website last updated?

- Within the past month
- 1-3 months ago
- 3-6 months ago
- 6-12 months ago
- Over a year ago / I don't know

6. When was your last email to your sphere?

- Within the past month

- 1-2 months ago
- 3-6 months ago
- 6+ months ago
- I don't email my sphere

Score Section 2:
Mostly "within the past month/week": Strong currency
Mostly 1-3 months: Moderate currency
Mostly 3+ months: Weak currency—you appear inactive

SECTION 3: SPECIFICITY (Are You Specific or Generic?)

7. If someone reads your bio, can they tell:

- What city/neighborhoods you serve
- What type of client you specialize in
- What you're known for (your edge)
- None of the above—it's generic

8. Does your content reference:

- Specific neighborhoods or streets
- Local businesses, schools, or landmarks
- Market trends specific to your area
- None of the above—it could apply anywhere

9. When someone looks at your last 10 social posts, do they see:

- Personality and voice (sounds like a real person)
- Local context (proof you live/work in this market)
- Value (useful information, not just announcements)
- Mostly generic content (could belong to any agent)

Score Section 3:
6+ checkmarks: Strong specificity
3-5 checkmarks: Moderate specificity
0-2 checkmarks: Weak specificity—you blend in

SECTION 4: USEFULNESS (Do You Teach or Just Announce?)

10. In the past month, have you:

- Explained a market trend in your own words
- Answered a common client question publicly
- Shared local insight (neighborhood, schools, development)
- Provided value unrelated to buying/selling (restaurant rec, community event)
- None of the above

11. When clients describe you to their friends, would they say:

- "They explain things really clearly"
- "They really know the market"
- "They're helpful and responsive"
- "I'm not sure what they'd say"

12. Does your content feel like:

- Teaching (helping people understand)
- Selling (promoting yourself or listings)
- Announcing (just stating facts)
- Mixed

Score Section 4:
Mostly teaching/helpful: Strong usefulness
Mostly mixed: Moderate usefulness

Mostly selling/announcing: Weak usefulness—you're not building trust

SECTION 5: PROOF (Do You Have Social Proof?)

13. Do you have visible reviews?

- Yes, 10+ recent reviews on Google
- Yes, 5-10 reviews
- Yes, fewer than 5 reviews
- No visible reviews

14. Do you have testimonials on your website?

- Yes, with specific details (names, outcomes, stories)
- Yes, but generic
- No

15. Can someone see evidence of recent closings?

- Yes, on social media or website (Just Sold posts, client stories)
- Occasionally
- No

Score Section 5:
Strong proof across all three: High social proof
Some proof in 1-2 areas: Moderate social proof
Little to no proof: Weak social proof—claims aren't validated

OVERALL TRUST AUDIT SCORE

Count your totals:

- Strong in 4-5 sections: Your presence confirms trust effectively

- Moderate in 3+ sections: You're visible but not fully credible—opportunities for improvement
- Weak in 3+ sections: You have credibility leaks—clients are hesitating when they check you

ACTION ITEMS (Based on Your Results)

If you scored WEAK in Findability:

- Claim and optimize your Google Business Profile
- Ensure your social profiles clearly state: Name + Realtor + City
- Update your website with accurate contact info

If you scored WEAK in Currency:

- Commit to Minimum Viable Presence: 1 email/month, 2-3 social posts/week
- Batch content in advance so you don't go dark during busy seasons
- Update your website quarterly minimum

If you scored WEAK in Specificity:

- Rewrite your bio to include: who you serve, where, and what you're known for
- Add local references to your next 5 posts (neighborhoods, streets, businesses)
- Create content that only you could create (based on your market knowledge)

If you scored WEAK in Usefulness:

- Shift from announcing to teaching: explain one thing per week
- Answer a common client question publicly

- Share value unrelated to transactions (community recommendations)

If you scored WEAK in Proof:

- Ask your last 5 happy clients for Google reviews
- Add 3-5 testimonials to your website (with specifics: names, outcomes)
- Post proof of recent activity (closings, client wins) on social media

This audit is designed to be repeated quarterly. Track your progress. Celebrate improvements. Stay honest about gaps.

APPENDIX B

The 70-20-10 Content Framework

Purpose: Eliminate guesswork about what to post. Build trust by providing value 90% of the time.

THE FRAMEWORK

70% — Value & Community
Content that has nothing to do with real estate transactions but builds connection and usefulness.

20% — Education & Authority
Content that demonstrates your expertise without being salesy.

10% — Direct Business
Content that promotes your listings, services, or asks for business.

70% CONTENT: VALUE & COMMUNITY

Goal: Be genuinely useful. Build affinity. Show personality.

Examples:

- Local restaurant or coffee shop recommendation
- Weekend activity guide ("5 things to do in [City] this weekend")

- Hidden gem spotlight (park, trail, venue)
- Community event announcement (farmer's market, festival, concert)
- Local business shoutout (support small businesses you love)
- Personal story (family moment, hobby, lesson learned)
- Seasonal tips (holiday events, summer activities, fall traditions)
- "Things I love about [Neighborhood]" post
- Behind-the-scenes of your life (not real estate—just you being human)

Mindset: How can I be helpful to my community, regardless of whether they ever hire me?

20% CONTENT: EDUCATION & AUTHORITY

Goal: Demonstrate competence. Reduce uncertainty. Teach.

Examples:

- Market insight explained simply ("Here's what the latest inventory data means for buyers")
- Common question answered ("Should we wait for rates to drop?")
- Myth-busting ("No, you don't need 20% down to buy a home")
- Neighborhood comparison ("East vs. West [City]: What's the difference?")
- Process education ("What happens between offer acceptance and closing?")

- Client success story (testimonial or case study)
- "What I'm seeing in the market right now" update
- Explanation of a headline ("What this new lending rule means for you")
- Video walkthrough of a neighborhood (showing features, explaining value)

Mindset: What do my clients misunderstand or worry about? How can I make it clearer?

10% CONTENT: DIRECT BUSINESS

Goal: Promote listings, services, open houses, or ask for referrals—sparingly.

Examples:

- New listing announcement
- Just sold celebration
- Open house invitation
- Client testimonial with a CTA ("If you or someone you know needs help, let's talk")
- "I'm currently helping clients in [neighborhood]—if you know anyone looking, send them my way"
- Year-end recap ("Here's what we accomplished this year—grateful for every client")

Mindset: I've provided massive value 90% of the time. It's okay to occasionally ask.

SAMPLE WEEKLY CONTENT CALENDAR (3 Posts)

Monday (70% - Value):
"5 coffee shops in [City] you need to try. My personal

favorite is [name] for their cold brew and the vibe." [Photo of coffee shop]

Wednesday (20% - Education):
"A lot of buyers are asking: 'Should we wait for rates to drop?' Here's how I think about it: [2-3 paragraphs explaining tradeoffs]. What questions do you have?" [Carousel or video]

Friday (70% - Community):
"This weekend: [City] Farmer's Market is back, live music at [venue], and the weather is perfect. What are you doing this weekend?" [Community photo]

Next week rotation might include one 10% post (listing or testimonial).

CONTENT BATCHING USING 70-20-10

Monthly Batching Session (2-3 hours):

70% Content (9-10 posts for the month):

- Write/capture 3 local recommendations
- Write/capture 3 community or personal posts
- Write/capture 3 weekend activity ideas

20% Content (3-4 posts for the month):

- Draft 1 market update
- Answer 1 common question
- Share 1 client success story or educational tip

10% Content (1-2 posts for the month):

- Draft 1 listing announcement or testimonial CTA

Result: 12-16 posts batched in advance. Schedule them. Stay consistent.

WHY THIS WORKS

Builds trust through reciprocity: You give value first, repeatedly. When you ask for business, people want to reciprocate.

Reduces algorithm pressure: You're not fighting for attention with salesy content. You're earning it with usefulness.

Differentiates you: Most agents post 80% business content (listings, sold signs, "I'm the best"). You're providing value 90% of the time. You stand out.

Sustainable: It's easier to stay consistent when 70% of your content is simple, human, and helpful.

APPENDIX C

The AI Context Packet: Teaching AI to Sound Like You

AI doesn't replace your judgment.
It reflects it.

When agents get poor results from AI, it's rarely because the tool is weak. It's because the inputs are vague, inconsistent, or misaligned with how they actually work with clients.

The solution isn't more prompting.
It's clear context.

We call this a Context Packet—a simple, one-time framework that teaches AI how you think, how you communicate, and where your boundaries are. When used correctly, it allows AI to produce drafts that sound like you, respect your standards, and require far less editing.

This appendix explains the structure of a strong Context Packet and how to use it effectively. A ready-to-use template is available at therealtorlab.com/edge.

1. Positioning: Clarifying Your Lane

Before AI can sound like you, it needs to know who you actually serve and where you add the most value.

A strong Context Packet clearly defines:

- The types of clients you work with most often
- The markets or neighborhoods you focus on
- What you are known for when clients describe you to others
- The specific lanes you operate in (two or three at most)

This isn't about sounding impressive.
It's about being accurate.

When AI understands your positioning, it stops producing generic content and starts reinforcing your expertise.

2. Voice and Tone: Protecting How You Sound

Most agents don't struggle with what to say.
They struggle with how it comes across.

Your Context Packet should describe:

- How you sound at your best
- The words and phrases that feel natural to you
- The language you intentionally avoid
- Your preferences around professionalism, warmth, clarity, and directness

This is where you protect yourself from hype, forced urgency, and sales-heavy language that erodes trust.

AI doesn't need personality.
It needs guardrails.

3. Communication Philosophy: How You Handle Uncertainty

Trust is built in moments of ambiguity.

Your Context Packet should explain:

- How you talk about uncertainty without inflating fear
- How you frame tradeoffs and decisions
- How you teach clients without overwhelming them
- What clients consistently say they appreciate about working with you

This teaches AI to mirror your judgment, not just your tone.

4. Content Principles: What Your Content Is For

Not all content serves the same purpose.

Clear principles help AI understand what your content should do, such as:

- Reduce uncertainty
- Demonstrate local expertise
- Teach without selling
- Build trust through usefulness
- Show humanity without oversharing

It should also clarify the topics you speak about often—and the ones you avoid entirely.

Boundaries create consistency.
Consistency creates trust.

5. Boundaries and Guardrails: What AI Must Never Do

This section protects both you and your clients.

A strong Context Packet explicitly states:

- What AI should never generate (pricing advice, legal guidance, invented data, risky language)
- How it should handle uncertainty
- When it should ask clarifying questions
- How calls-to-action should feel (service-oriented, not sales-driven)

This is where AI stops being a risk and starts being a reliable assistant.

6. Examples: Teaching by Showing

Finally, the most important section: examples of your best work.

When AI can see what good looks like, everything improves.

Strong examples share:

- Clear structure
- Calm, confident tone
- Market-specific insight
- Value without pressure

These examples anchor AI to your real voice—not an abstract description of it.

Using the Context Packet

Once created, your Context Packet becomes the foundation for all AI-assisted work.

You don't rewrite it every time.

You reference it.

A complete, fillable template—and examples of how to use it with AI—are available at therealtorlab.com/edge.

APPENDIX D

Monthly Email Frameworks That Reinforce Trust

Email remains one of the most reliable trust-building tools in real estate—when it's done well.

The goal of monthly emails isn't volume.

It's presence.

This appendix outlines four proven email frameworks that help you stay relevant, useful, and human without sounding repetitive or sales-driven. Customizable versions of these templates are available at therealtorlab.com/edge.

1. Market Update Emails

Purpose: Reduce uncertainty by explaining what's happening and why it matters.

Effective market updates:

- Focus on trends, not predictions
- Translate data into implications
- Avoid hype and alarmism
- End with availability, not pressure

These emails position you as a steady interpreter of change.

2. Neighborhood Spotlight Emails

Purpose: Demonstrate local expertise.

Strong neighborhood spotlights explain:

- What's changing
- Why people choose the area
- Who it's a good fit for
- How the market behaves there

This is teaching, not promotion.

3. Helpful Resource Emails

Purpose: Teach without selling.

These emails address common questions or decisions clients face and break them down clearly:

- What to expect
- What matters most
- What people often misunderstand

They build trust by being useful—whether or not someone is transacting.

4. Personal Update Emails

Purpose: Reinforce relationship.

The best personal updates:

- Share small, genuine moments

- Connect business to human experience
- Express gratitude without obligation

These emails remind people there's a real person behind the expertise.

Using These Frameworks

Rotate through these formats to stay consistent without becoming predictable.

A simple rhythm works:

- One email per month minimum
- Vary the framework
- Keep the tone calm, clear, and human

AI can help draft these emails—but only when guided by your Context Packet.

Both tools, used together, create leverage without sacrificing trust.

END OF APPENDICES

Final Notes to the Reader

You've reached the end of The Realtor Edge.

If you've read this far, you're not a casual observer. You're someone who cares about building something durable.

The concepts in this book aren't theoretical. They're field-tested, practiced daily, and proven across decades and market cycles by agents who refused to compete on hype and chose instead to compete on trust.

Now it's your turn.

Start with the 90-Day Plan. Build your foundation. Master one channel. Stay consistent. Refine as you go.

And a year from now—when your pipeline is full, your sphere is engaged, and referrals are flowing—you'll understand what we mean by the Realtor Edge.

It's not loud.

It's not flashy.

It's just durable.

And durable wins.

For tools, templates, and ongoing support:
Visit therealtorlab.com/edge

Thank you for reading. Now go build something that lasts.

— Stan & Chase

Made in United States
Troutdale, OR
11/03/2026

56845382R00156